S

Exponential Time Algorithms

Serge Gaspers

# Exponential Time Algorithms

## Structures, Measures, and Bounds

VDM Verlag Dr. Müller

## Impressum/Imprint (nur für Deutschland/ only for Germany)

Bibliografische Information der Deutschen Nationalbibliothek: Die Deutsche Nationalbibliothek verzeichnet diese Publikation in der Deutschen Nationalbibliografie; detaillierte bibliografische Daten sind im Internet über http://dnb.d-nb.de abrufbar.

Alle in diesem Buch genannten Marken und Produktnamen unterliegen warenzeichen-, marken- oder patentrechtlichem Schutz bzw. sind Warenzeichen oder eingetragene Warenzeichen der jeweiligen Inhaber. Die Wiedergabe von Marken, Produktnamen, Gebrauchsnamen, Handelsnamen, Warenbezeichnungen u.s.w. in diesem Werk berechtigt auch ohne besondere Kennzeichnung nicht zu der Annahme, dass solche Namen im Sinne der Warenzeichen- und Markenschutzgesetzgebung als frei zu betrachten wären und daher von jedermann benutzt werden dürften.

Coverbild: www.purestockx.com

Verlag: VDM Verlag Dr. Müller Aktiengesellschaft & Co. KG
Dudweiler Landstr. 99, 66123 Saarbrücken, Deutschland
Telefon +49 681 9100-698, Telefax +49 681 9100-988, Email: info@vdm-verlag.de
Zugl.: Bergen, University of Bergen, Diss., 2008

Herstellung in Deutschland:
Schaltungsdienst Lange o.H.G., Berlin
Books on Demand GmbH, Norderstedt
Reha GmbH, Saarbrücken
Amazon Distribution GmbH, Leipzig
**ISBN: 978-3-639-21825-1**

## Imprint (only for USA, GB)

Bibliographic information published by the Deutsche Nationalbibliothek: The Deutsche Nationalbibliothek lists this publication in the Deutsche Nationalbibliografie; detailed bibliographic data are available in the Internet at http://dnb.d-nb.de .

Any brand names and product names mentioned in this book are subject to trademark, brand or patent protection and are trademarks or registered trademarks of their respective holders. The use of brand names, product names, common names, trade names, product descriptions etc. even without a particular marking in this works is in no way to be construed to mean that such names may be regarded as unrestricted in respect of trademark and brand protection legislation and could thus be used by anyone.

Cover image: www.purestockx.com

Publisher:
VDM Verlag Dr. Müller Aktiengesellschaft & Co. KG
Dudweiler Landstr. 99, 66123 Saarbrücken, Germany
Phone +49 681 9100-698, Fax +49 681 9100-988, Email: info@vdm-publishing.com

Printed in the U.S.A.
Printed in the U.K. by (see last page)
**ISBN: 978-3-639-21825-1**

# Preface

Although I am the author of this book, I cannot take all the credit for this work.

## Publications

This book is a revised and updated version of my PhD thesis. It is partly based on the following papers in conference proceedings or journals.

**[FGPR08]** F. V. Fomin, S. Gaspers, A. V. Pyatkin, and I. Razgon, *On the minimum feedback vertex set problem: Exact and enumeration algorithms*, Algorithmica **52**(2) (2008), 293–307. A preliminary version appeared in the proceedings of IWPEC 2006 [FGP06].

**[GKL08]** S. Gaspers, D. Kratsch, and M. Liedloff, *On independent sets and bicliques in graphs*, Proceedings of WG 2008, Springer LNCS 5344, Berlin, 2008, pp. 171–182.

**[GS09]** S. Gaspers and G. B. Sorkin, *A universally fastest algorithm for Max 2-Sat, Max 2-CSP, and everything in between*, Proceedings of SODA 2009, ACM and SIAM, 2009, pp. 606–615.

**[GKLT09]** S. Gaspers, D. Kratsch, M. Liedloff, and I. Todinca, *Exponential time algorithms for the minimum dominating set problem on some graph classes*, ACM Transactions on Algorithms **6**(1:9) (2009), 1–21. A preliminary version appeared in the proceedings of SWAT 2006 [GKL06].

**[FGS07]** F. V. Fomin, S. Gaspers, and S. Saurabh, *Improved exact algorithms for counting 3- and 4-colorings*, Proceedings of COCOON 2007, Springer LNCS 4598, Berlin, 2007, pp. 65–74.

**[FGSS09]** F. V. Fomin, S. Gaspers, S. Saurabh, and A. A. Stepanov, *On two techniques of combining branching and treewidth*, Algorithmica **54**(2) (2009), 181–207. A preliminary version received the Best Student Paper award at ISAAC 2006 [FGS06].

**[FGK+08]** F. V. Fomin, S. Gaspers, D. Kratsch, M. Liedloff, and S. Saurabh, *Iterative compression and exact algorithms*, Proceedings of MFCS 2008, Springer LNCS 5162, Berlin, 2008, pp. 335–346.

# Acknowledgments

First and foremost, I would like to express my gratitude to my PhD supervisor Fedor V. Fomin. His guidance, insight, and knowledge were of great help for my intellectual advancement. Ideas were always much clearer when leaving his office than they were before.

I would also like to sincerely thank my co–supervisor Pınar Heggernes for being a great teacher, an instructive researcher and a welcoming person. I would also like to thank the members of my PhD committee, Dag Haugland, Thore Husfeldt, and Ryan Williams. My deepest thanks go to my Master thesis supervisor, Dieter Kratsch, whom I owe my interest in algorithms, who inspired me, taught me, and introduced me to the area of exponential time algorithms.

Many thanks for collaboration, fruitful discussions, inspiring ideas, and teaching me valuable things go to Jérémy Barbay, Stéphane Bessy, Binh-Minh Bui-Xuan, Bruno Courcelle, Jean Daligault, Frederic Dorn, Michael R. Fellows, Henning Fernau, Stephen Finbow, Martin Fürer, Petr Golovach, Fabrizio Grandoni, Sylvain Guillemot, Mamadou M. Kanté, Shiva P. Kasiviswanathan, Marcos Kiwi, William F. Klostermeyer, Jan Kratochvíl, Alexander S. Kulikov, Konstantin Kutzkov, Mathieu Liedloff, Daniel Lokshtanov, Elena Losievskaja, Benoît Martin, Daniel Meister, Margaret-Ellen Messinger, Rodica Mihai, Matthias Mnich, Jesper Nederlof, Richard J. Nowakowski, Paul Ottaway, Christophe Paul, Anthony Perez, Paweł Prałat, Artem V. Pyatkin, Daniel Raible, Michaël Rao, Ivan Rapaport, Igor Razgon, Frances A. Rosamond, Peter Rossmanith, Saket Saurabh, Alexander D. Scott, Gregory B. Sorkin, Maya J. Stein, Alexey A. Stepanov, Karol Suchan, Jan Arne Telle, Stéphan Thomassé, Ioan Todinca, Yngve Villanger, Magnus Wahlström, David Wolfe, David R. Wood, Paul Yans, and Norbert Zeh.

For financial support, I would like to thank the Norwegian Research Council (NFR), the French National Research Agency (ANR), the L. Meltzers Høyskolefond, the Center for Mathematical Modeling, and the Universities of Bergen (Norway), Montpellier 2 (France), and Chile.

On the personal side, I would like to thank my friends and my family, and especially my parents Esther and Guy, my brother Sven, and my grandmother Mathilde, who sadly died in the beginning of 2008 and to whom I dedicate this thesis. For all her love, support and great company I would like to thank my girlfriend Nancy — I love you.

*Fir méng Bomi.*

# Contents

# List of Figures

# List of Tables

# Chapter 1

# Introduction

One does not fear the perebor[1] but rather uses
it reasonably via a realistic estimation of the
dimensions of the disaster it may imply.

*Adel'son-Vel'skii et al.* [AVAD76]
(translated by Trakhtenbrot [Tra84])

"Eureka - You Shrink!" shouted out 28-year old Jack Edmonds in 1963 when he discovered new insights into the MATCHING problem, says a legend.[2] These insights were fundamental for his polynomial time MATCHING algorithm. In the paper [Edm65] presenting this algorithm he distinguished *good* algorithms from *bad* algorithms, that is those requiring only polynomial computation time in the length of the input from those that are subject to the 'curse of exponentiality'. Before, researchers were mainly distinguishing between finite and infinite computation — no wonder, because one of the only so far implemented algorithms for combinatorial optimization problems was the `Simplex` algorithm which has worst case exponential running time, but performed well in practice nevertheless.

As computing devices became more popular, more and more people experienced that good algorithms were indeed fast and that exponential time algorithms were slow on large inputs.

By the end of the '60s it became ever more clear that there were problems resisting good algorithms to solve them. This led Stephen A. Cook to define the class $\mathcal{NP}$, of problems for which a solution (a certificate) can be checked in time polynomial in the instance size. It is clear that the class $\mathcal{P}$, of problems that can be solved in polynomial time, is a subset of $\mathcal{NP}$, but whether or not it is a proper subset of $\mathcal{NP}$ is not known until today. The famous $\mathcal{P}$ vs. $\mathcal{NP}$ question is nowadays one of the most important open questions in science. In his influential paper [Coo71] introducing the class $\mathcal{NP}$, Cook also proves that the SAT problem belongs to the class of $\mathcal{NP}$-complete problems, the subset of the hardest problems in $\mathcal{NP}$. A polynomial time solution to one $\mathcal{NP}$-complete problem would imply that $\mathcal{P} = \mathcal{NP}$. Leonid A.

---

[1]Russian for "brute force" or "exhaustive search", but more accurately refers here to any exponential time search procedure

[2]Edmonds confirmed this at the Aussois 2001 workshop [JRR03], except that instead of 'Eureka', it maybe was some less dignified word

Levin established the same result independently [Lev73]. In the year following the publication of Cook's $\mathcal{NP}$-completeness theorem, Richard M. Karp [Kar72] proposed a general method for proving the $\mathcal{NP}$-completeness of other problems by reductions to problems already known to be $\mathcal{NP}$-complete, and proves $\mathcal{NP}$-completeness for another 21 well known problems. At the end of the '70s, Garey and Johnson [GJ79] published the first book on $\mathcal{NP}$-completeness which still serves as a reference to many researchers. Besides presenting $\mathcal{NP}$-completeness results for many problems, the book also raises the question of how to cope with intractable problems.

As defined above, many problems, like MAXIMUM INDEPENDENT SET[3], are not in $\mathcal{NP}$ unless $\mathcal{P} = \mathcal{NP}$. Given a graph $G$ and an independent set of $G$, it cannot be checked in polynomial time whether this is a maximum independent set for $G$ unless $\mathcal{P} = \mathcal{NP}$. The crux is here that the class $\mathcal{NP}$ has been designed for *decision* problems, such as the following: given a graph $G$ and a constant $k$, does $G$ admit an independent set of size at least $k$? This problem can be shown to be $\mathcal{NP}$-complete [Kar72]. For *optimization* versions of the problems one usually speaks of $\mathcal{NP}$-hard problems if the corresponding decision version is $\mathcal{NP}$-complete. More formally, a problem $P_1$ is $\mathcal{NP}$-hard if an algorithm (or oracle) solving $P_1$ in polynomial time makes it possible to solve an $\mathcal{NP}$-complete problem $P_2$ in polynomial time, that is $P_2$ can be reduced to $P_1$ in polynomial time. Other kinds of problems often considered are *counting* problems, where one is asked to count all objects respecting certain criteria, and *enumeration* problems where one is asked to list all objects respecting certain criteria. Several complexity classes were introduced to capture the hardness of these problems: $\mathcal{NPO}$ for optimization problems [ACG+99], $\#\mathcal{P}$ for counting problems [Val79], and $\mathcal{ENP}$ and several subclasses for enumeration problems [Fuk96].

Various methods have, since the history of $\mathcal{NP}$-completeness, been studied to confront $\mathcal{NP}$-hard problems, such as heuristics, approximation algorithms, randomized algorithms, average case complexity, fixed parameter tractability, the restriction of the problems instances to classes of polynomial time solvable instances, hybrid algorithms (see [VWW06]), or the design of exponential time algorithms for moderate instance sizes.

This book is about one such strategy to deal with $\mathcal{NP}$-hard problems, namely exponential time algorithms. In this area of algorithm research, we try to design algorithms that *solve* hard problems *exactly*[4] and whose *worst case* running time is dominated by a "small"[5] *exponential* function of the input size. The main point is that a reasonable amount of exponentiality is the price we have to pay to obtain exact solutions to combinatorial hard problems, unless $\mathcal{P} = \mathcal{NP}$.

In the literature, many games and puzzles (or variants of them) have been shown to be $\mathcal{NP}$-complete, such as FreeCell, Mahjong, Mastermind, Minesweeper, Sudoku and Tetris; see [KPS08] for a recent survey. Nevertheless, they are routinely solved by humans and experienced players even crack the hardest instances in a reasonable amount of time. Even if there is no known polynomial time algorithm for these games and puzzles, humans can still solve them if

---

[3]words and symbols displayed in dark gray are defined in the glossary on page 177 or in the problem definition list on page 181

[4]in contrast to heuristics or approximation algorithms

[5]here, small is relative to the considered problem; for some problems, a running time of the form $\mathcal{O}(c^n)$ for any constant $c$ would be a great achievement, and for other problems running times of, say $\mathcal{O}(1.4^n)$, are easily obtained

the instances are not too large. The same is true for computer programs: on moderately sized instances, even exponential time algorithms perform fast.

As customary, for input instances that are graphs, formulas or sets, we denote by $n$ the number of vertices, variables or elements of the input and by $m$ the number of edges, clauses or sets of the input. Unless specified in the problem definition list on page 181 or otherwise, we denote by $n$ the length of the input for all other types of input instances. As we are mainly interested in the exponential part of the running time of algorithms, let us also use a modified "big-Oh" notation that suppresses polynomially bounded terms. For an exponential function $f$,

$$\mathcal{O}^*(f(n)) := f(n) \cdot n^{\mathcal{O}(1)}.$$

In the above definition and in the remainder of this text, we follow the commonly used convention that a function $g(n)$ containing a big-Oh term $\mathcal{O}(h(n))$ is the union of all $\mathcal{O}(g'(n))$ such that $g'(n)$ can be obtained from $g(n)$ by substituting $\mathcal{O}(h(n))$ with a function belonging to $\mathcal{O}(h(n))$. We also write $f(n) = \mathcal{O}(g(n))$ when we actually mean $f(n) \in \mathcal{O}(g(n))$.

Quite early, researchers in theoretical computer science realized that some $\mathcal{NP}$-hard problems can be solved provably faster than with a trivial brute-force search through the space of all candidate solutions. Early examples of exponential time algorithms include

- an $\mathcal{O}^*(2^n)$ algorithm for the Traveling Salesman problem by Held and Karp [HK62], which is still the fastest known algorithm for this problem,

- an $\mathcal{O}^*(2^{n/2})$ algorithm for the Binary Knapsack problem by Horowitz and Sahni [HS74],

- an $\mathcal{O}(2.4423^n)$ algorithm for the Chromatic Number problem and a $\mathcal{O}(1.4423^n)$ algorithm for the 3-Coloring problem by Lawler [Law76],

- an $\mathcal{O}(1.2599^n)$ algorithm for the Maximum Independent Set problem by Tarjan and Trojanowski [TT77],

- an $\mathcal{O}^*(2^{n/2})$ algorithm for Binary Knapsack and other problems by Schroeppel and Shamir [SS81], which uses less space ($\mathcal{O}^*(2^{n/4})$) than the one by Horowitz and Sahni,

- an $\mathcal{O}^*(2^n)$ polynomial space algorithm for the Hamiltonian Path problem by Karp [Kar82]

- an $\mathcal{O}(1.6181^n)$ algorithm for the 3-Sat problem by Monien and Speckenmeyer [MS85], and

- $\mathcal{O}(1.2346^n)$ and $\mathcal{O}(1.2109^n)$ algorithms for the Maximum Independent Set problem by Jian [Jia86] and Robson [Rob86]

Only scattered results (for example [ST90, GSB95, BE95, Zha96, Kul99, Bei99]) in the area of exponential time algorithms appeared in the literature in the '90s. Probably due to a DIMACS Workshop on "Faster Exact Solutions for $\mathcal{NP}$ Hard Problems" in the year 2000 and a seminal survey on exponential time algorithms by Woeginger [Woe03], the area of exponential time algorithms gained more and more interest in the theoretical computer science community. Many PhD theses [Bys04b, Gra04, Ang05, Rie06, Bjö07, Dor07, Lie07, Wah07, Wil07, Gas08, Ste08] have been completely or partly devoted to the topic of exponential time algorithms and several surveys [DHIV01, Woe03, Iwa04, Woe04, FGK05, Sch05, Woe08] on exponential time algorithms have appeared since then.

In the next section we survey some negative results that show limits of what we can reasonably hope to achieve in terms of running times of algorithms for $\mathcal{NP}$-hard problems. In Section 1.2 we give an overview of some important techniques used to design exponential time algorithms. Section 1.3 is a rather informal discussion on time and space complexities of algorithms. Finally, Section 1.4 gives an overview of the remaining chapters.

## 1.1 Negative Results

In this section, a few negative results are given. Some of these motivate the study of exponential time algorithms, others give reasonable limits of what can be achieved by exponential time algorithms.

Unless $\mathcal{P} = \mathcal{NP}$, there exists no polynomial time algorithm for $\mathcal{NP}$-hard problems. Moreover it is now widely believed that $\mathcal{P} \neq \mathcal{NP}$ [Gas02], which means that superpolynomial time algorithms are the best we can hope to achieve to solve $\mathcal{NP}$-hard problems.

Note that $\mathcal{P} \neq \mathcal{NP}$ would not imply that there exists no subexponential time algorithm for some hard problems. Here, subexponential time means time $2^{o(n)}$. In his thesis, Dorn [Dor07] studies several problems for which he exhibits subexponential time algorithms. These problems are often restricted to planar graphs, graphs with bounded genus or graphs excluding other graphs as minors. However, it is widely believed (see [Woe03]) that many $\mathcal{NP}$-hard problems are not solvable in subexponential time. More precisely, the Exponential Time Hypothesis, by Impagliazzo and Paturi [IP99], states the following.

**Conjecture** (Exponential Time Hypothesis [IP99]). *There is no subexponential time algorithm for* 3-SAT.

By reductions preserving subexponential time complexities, Impagliazzo et al. [IPZ01] proved that under the Exponential Time Hypothesis, several other problems, like 3-COLORING, HAMILTONIAN CYCLE and MAXIMUM INDEPENDENT SET do not have a subexponential time algorithm either. Johnson and Szegedy [JS99] even strengthen the result for MAXIMUM INDEPENDENT SET showing that the problem does not admit a subexponential time algorithm under the Exponential Time Hypothesis when restricted to graphs of maximum degree 3. For a number of other problems, it has been proved that they do not have subexponential time algorithms unless the Exponential Time Hypothesis fails.

Further, Pudlák and Impagliazzo [PI00] prove that for every $k \geq 3$ there exists a constant $c_k > 0$, where $c_k \to 0$ as $k \to \infty$, such that every DLL algorithm for $k$-SAT has running time $\Omega(2^{n \cdot (1 - c_k)})$. Here, a DLL algorithm is an algorithm that selects a variable $v$ at each step, and recursively solves subproblems where $v$ is set to true and false respectively, unless some clause is falsified, in which case it stops. Similarly, Traxler [Tra08] gives some evidence that no

$\mathcal{O}(c^n)$ algorithm exists for $(d, 2)$-CSP, namely that under the Exponential Time Hypothesis, any algorithm for $(d, 2)$-CSP has running time $\Omega(d^{c \cdot n})$ for some constant $c > 0$ independent of $d$.

Finally, showing the limits of approximation algorithms, many $\mathcal{NP}$-hard problems have been proved to be hard to approximate in polynomial time. For example, it is $\mathcal{NP}$-hard to approximate MAXIMUM INDEPENDENT SET within a factor of $n^{1-\varepsilon}$ for any $\varepsilon > 0$ [Zuc07].

## 1.2 Overview of Techniques

In this section, we describe some of the known techniques for designing and analyzing exponential time algorithms and give examples for some of the techniques. We focus on deterministic algorithms here.

### 1.2.1 Brute Force

Every problem in $\mathcal{NP}$ can be solved by exhaustive search over all candidate solutions. The search space, that is the set of all candidate solutions, has size

- $\mathcal{O}(2^n)$ for subset problems,

- $B_n = \mathcal{O}(c^{n \log n})$ for a constant $c > 1$ for partitioning problems where $B_n$ denotes the $n^{\text{th}}$ Bell number, and

- $\mathcal{O}(n!)$ for permutation problems,

where the ground set has size $n$. As for every problem in $\mathcal{NP}$, a candidate solution can be checked in polynomial time, the running time of a brute force algorithm is within a polynomial factor of the size of the search space.

Whereas for many hard problems, better running time bounds have been achieved,

- for subset problems like SAT[6], MINIMUM HITTING SET, and MINIMUM SET COVER, no algorithm of time complexity $\mathcal{O}(1.9999^n)$ is known,

- for partitioning problems like CSP and GRAPH HOMOMORPHISM no known algorithm has time complexity $\mathcal{O}(c^n)$ for a constant $c$, and

- for permutation problems like QUADRATIC ASSIGNMENT and SUBGRAPH ISOMORPHISM, no known algorithm has time complexity $\mathcal{O}(c^n)$ for a constant $c$.

---

[6]The currently fastest deterministic algorithm for SAT is due to Dantsin et al. [DHW06] and has running time $\mathcal{O}^* \left( 2^{n \cdot \left(1 - \frac{1}{\ln(m/n) + \mathcal{O}(\ln \ln m)}\right)} \right)$ .

### 1.2.2   Bounds on Mathematical Objects

A very natural question in graph theory is: how many minimal (maximal) vertex subsets satisfying a given property can be contained in a graph on $n$ vertices? The trivial bound is $\mathcal{O}\left(\binom{n}{n/2}\right)$, which is $\mathcal{O}(2^n/\sqrt{n})$ by Stirling's approximation. Only for few problems better bounds, that is bounds of the form $\mathcal{O}(c^n)$ for $c < 2$, are known. One example of such a bound is the celebrated Moon and Moser [MM65] theorem, basically stating that every graph on $n$ vertices has at most $3^{n/3}$ maximal cliques (independent sets). Another example is the result from [FGPS08], where it is shown that the number of minimal dominating sets is at most $1.7170^n$.

Besides their combinatorial interest, such bounds often have algorithmic consequences. Worst–case upper bounds on the (total) running time of enumeration algorithms can be proved using these bounds. For example, the Moon and Moser theorem implies an overall running time of $\mathcal{O}^*(3^{n/3}) = \mathcal{O}(1.4423^n)$ for the polynomial delay algorithm of Johnson et al. [JYP88] for enumerating all maximal independent sets of a graph on $n$ vertices.

More indirect consequences are worst case upper bounds on the running time of algorithms using an enumeration algorithm as a subroutine, such as many COLORING algorithms [Law76, Epp03, Bys04a, Bys04b, BH08] or algorithms for other problems that enumerate maximal independent sets [RSS07].

The simplest such example is probably Lawler's algorithm [Law76] for the $\mathcal{NP}$–hard problem to decide whether a graph $G$ is 3-colorable. The algorithm works as follows: for each maximal independent set $I$ of $G$, check if $G\backslash I$ is bipartite. If for at least one maximal independent set $I$, $G \setminus I$ is bipartite then $G$ can be colored with 3 colors. As deciding if a graph is bipartite is polynomial time solvable, the algorithm has time complexity $\mathcal{O}^*(3^{n/3})$.

Upper bounds of $\mathcal{O}(1.6181^n)$ for the number of minimal separators and $\mathcal{O}(1.7347^n)$ for the number of potential maximal cliques in a graph have been proved and used by Fomin and Villanger [FV08, FV10] to design the currently fastest algorithms for the TREEWIDTH and the FEEDBACK VERTEX SET problem, amongst others. In [Moo71, GM09], the number of minimal feedback vertex sets in tournaments has been studied. Other results include bounds on the number of maximal induced $r$-regular subgraphs [GRS06] and on the number of maximal induced bipartite graphs [BMS05].

In Chapter 3, an upper bound of $1.8638^n$ on the number of minimal feedback vertex sets is presented and in Chapter 4 we show an upper bound of $n \cdot 3^{n/3}$ on the number of maximal bicliques, which is tight up to a linear factor.

### 1.2.3   Dynamic Programming Across Subsets

The idea of this technique is to store, for each subset of a ground set on $n$ elements (and often some additional information), a partial solution to the problem in an exponential size table so that the partial solutions can be looked up quickly. Dynamic programming across subsets can be used whenever a solution to an instance can be extended in polynomial time based on the solutions to all subinstances no matter how the solutions to the subinstances were obtained.

Consider the TRAVELING SALESMAN problem.

TRAVELING SALESMAN: Given a set $\{1, \dots, n\}$ of $n$ cities and the distance $d(i, j)$ between every two cities $i$ and $j$, find a tour visiting all cities with minimum total distance. A *tour* is a permutation of the cities starting and ending in city 1.

The $\mathcal{O}^*(2^n)$ algorithm for the TRAVELING SALESMAN problem by Held and Karp [HK62] works as follows. For a subset $S \subseteq \{2, \dots, n\}$ of cities and a city $i \in S$, let $\mathsf{Opt}[S, i]$ denote the length of the shortest path starting in city 1, visiting all cities in $S \setminus \{i\}$ and ending in city $i$. Then,

$$\mathsf{Opt}[S, i] := \begin{cases} d(1, i) & \text{if } S = \{i\}, \text{ and} \\ \min_{j \in S \setminus \{i\}} \{\mathsf{Opt}[S \setminus \{i\}, j] + d(i, j)\} & \text{otherwise.} \end{cases}$$

It is then straightforward to compute $\mathsf{Opt}[S, i]$ for each $S \subseteq \{2, \dots, n\}$ and $i \in S$ by going through the subsets in increasing order of cardinality. The time spent to compute the entry for one couple $(S, i)$ is polynomial by looking up the values $\mathsf{Opt}[S \setminus \{i\}, j]$ for each $j \in S \setminus \{i\}$. The final solution to the problem is then $\min_{j \in \{2, \dots, n\}} \{\mathsf{Opt}[\{2, \dots, n\}, j] + d(j, 1)\}$. Despite its simplicity, this algorithm is still the fastest known for the TRAVELING SALESMAN problem.

For a variant of the TRAVELING SALESMAN problem where the instance is given by a graph $G$ with maximum degree $d = \mathcal{O}(1)$ where the vertices represent cities and weighted edges represent the distance between these cities, Björklund et al. [BHKK08b] use a slight modification of the former algorithm to solve this variant in time $\mathcal{O}((2 - \varepsilon_d)^n)$ where $\varepsilon_d > 0$ depends on $d$ alone. Dynamic Programming has also been used to derive the $\mathcal{O}(2.4423^n)$ algorithm for the COLORING problem by Lawler [Law76], which was the fastest known algorithm for this problem for 25 years.

## 1.2.4 Branching

A branching algorithm selects at each step a local configuration of the instance and recursively solves subinstances based on all possible values this local configuration can take. This technique is presented in detail in Chapter 2 and is at the heart of many of the fastest known algorithms for various hard problems.

## 1.2.5 Memorization

Memorization was introduced by Robson [Rob86] as a trade-off between time and space usage. The idea is to precompute the solutions to subinstances of small size, say of size at most $\alpha n$, $\alpha < 1$, by Dynamic Programming and to look them up whenever a branching algorithm encounters a subinstance of size at most $\alpha n$, thereby reducing its running time. Alternatively, the solutions to small subinstances can also be computed on the fly by the branching algorithm and looked up whenever a solution to the subproblem has already been computed. We refer to [Rob86, Gra04, FGK05] for more details on the technique.

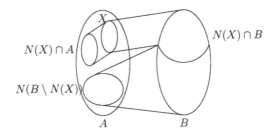

Figure 1.1: Illustration of the main phase of Liedloff's algorithm for DOMINATING SET in bipartite graphs

### 1.2.6   Preprocessing

The idea of this approach is to first perform an initial analysis or restructuring of the given input such that later on, queries to the precomputed values can be answered quickly. Preprocessing an exponentially large data set may lead to an exponential speedup in the running time of the algorithm. As an example, consider the following outline of Liedloff's algorithm [Lie08] for finding a minimum dominating set in a bipartite graph $G = (A, B, E)$ in time $\mathcal{O}^*(2^{n/2})$.

Let $B = \{b_1, \ldots, b_{|B|}\}$ be the largest of the two sets $A$ and $B$. The algorithm performs two phases, each having running time $2^{|A|} \cdot n^{\mathcal{O}(1)}$.
In the preprocessing phase, compute for each subset $X \subseteq A$ and integer $k$, $1 \leq k \leq |B|$, a subset $\mathsf{Opt}[X, k]$ which is a smallest subset of $\{b_1, \ldots, b_k\}$ that dominates $X$. This phase can be performed by Dynamic Programming in time $2^{|A|} \cdot n^{\mathcal{O}(1)}$.
In its main phase (see Figure 1.1), the algorithm goes again through all subsets $X \subseteq A$ and for each set $X$, it computes a dominating set $D$ of $G$ of minimum size such that $D \cap A = X$. For each set $X \subseteq A$, such a dominating set $D$ can be obtained by setting

$$D := X \cup (B \setminus N(X)) \cup \mathsf{Opt}[A \setminus (N[X] \cup N(B \setminus N(X))), |B|].$$

Observe that $X \subseteq D$ by definition, the vertices in $B \setminus N(X)$ must be added to $D$ to dominate themselves (since $B$ is an independent set), and, additionally, a minimum sized subset of vertices of $B$ dominating the vertices in $A \setminus (N[X] \cup N(B \setminus N(X)))$ (the only vertices that are not dominated by $X$ nor $B \setminus N(X)$) can be looked up in $\mathsf{Opt}[A \setminus (N[X] \cup N(B \setminus N(X))), |B|]$. This phase of the algorithm also takes time $2^{|A|} \cdot n^{\mathcal{O}(1)}$.

Preprocessing is a fundamental tool in algorithm design. For exponential time algorithms, it has been applied to BINARY KNAPSACK [HS74, SS81], EXACT HITTING SET [DP02], SUBSET SUM [Woe03], and several other problems [KW05, FGK+07]. The algorithms of Feige [Fei00] and Cygan and Pilipczuk [CP08, CP09a] for the BANDWIDTH problem could also be classified under Preprocessing algorithms.

### 1.2.7   Local Search

Local search algorithms explore the search space of candidate solutions by moving from one candidate solution to the next in a local fashion. This method has mainly be employed to design algorithms for $k$-SAT; see [Sch01] for an overview of local search algorithms.

Consider the 3-SAT problem. The space of candidate solutions is $\{0,1\}^n$, that is all possible truth assignments for the $n$ variables of a logical formula in conjunctive normal form. For a truth assignment $t \in \{0,1\}^n$ and a distance $d$, let $\mathcal{H}(t,d)$ denote the Hamming Ball of radius $d$ around $t$, that is the set of all truth assignments with Hamming distance at most $d$ from $t$.

An $\mathcal{O}^*(3^d)$ algorithm to check whether there exists a satisfying assignment in $\mathcal{H}(t,d)$ can easily be obtained as follows. As long as the current assignment is not a satisfying assignment, choose an unsatisfied clause, and go over all (at most 3) possibilities to flip the truth assignment of a variable in this clause and recurse with the new truth assignment and $d-1$.

Using a simple, so–called covering–code, of only the two Hamming balls $\mathcal{H}(\{0\}^n, \lfloor n/2 \rfloor)$ and $\mathcal{H}(\{1\}^n, \lceil n/2 \rceil)$, the whole search space is clearly covered. Applying the local search algorithm of the previous paragraph to both Hamming balls, it is easy to solve 3-SAT in time $\mathcal{O}^*(3^{n/2}) = \mathcal{O}(1.7321^n)$. More involved choices for the covering–codes and a slightly faster local search procedure inside the Hamming balls have led to the currently fastest deterministic algorithms for 3-SAT [DGH+02, BK04, Sch08].

### 1.2.8   Split and List

The Split and List method, which is quite similar to a preprocessing technique used in [HS74] and [SS81], is described and used by Williams [Wil05] to obtain an $\mathcal{O}^*(2^{\omega n/3})$ algorithm for MAX 2-CSP, where $\omega < 2.376$ is the matrix multiplication exponent. The basic idea is to split the variables into $k \geq 3$ equal sized parts, list all assignments of the variables in each of the parts, and finally combine all the solutions with a polynomial time algorithm on an exponentially large instance.

A Split and List algorithm for the MAX CUT problem works as follows [Wil05]. Divide the vertices of the graph $G = (V, E)$ in three parts $P_0, P_1, P_2$ of size roughly $n/3$ each. Build a complete 3-partite auxiliary graph containing a vertex for each subset of $P_0, P_1$ and $P_2$. Given vertices $x_i$ and $x_j$, $i \in \{0,1,2\}$, $j = i+1 \pmod 3$, that correspond to subsets $X_i, X_j$ of $P_i, P_j$ respectively, set the weight $w(x_i x_j)$ of the edge $x_i x_j$ to be the number of edges between $X_i \cup X_j$ and $P_i \setminus X_i$ plus the number of edges between $X_i$ and $P_j \setminus X_j$, that is

$$w(x_i x_j) := |N(X_i) \cap P_i| + |N(X_j) \cap (P_i \setminus X_i)| + |N(X_i) \cap (P_j \setminus X_j)|.$$

Then the weight of a triangle $x_1 x_2 x_3$ in the auxiliary graph corresponds to the number of edges that cross the partition of vertices $(V_l, V \setminus V_l)$ where $V_l = X_1 \cup X_2 \cup X_3$. To determine if $G$ contains a cut with $k$ edges, go over all $\mathcal{O}(m^3)$ possible triples $(k_{01}, k_{12}, k_{20})$ such that $k = k_{01} + k_{12} + k_{20}$, keep only those edges $x_i x_j$ in the auxiliary graph that have $w(x_i x_j) = k_{ij}$, and find a triangle in the auxiliary graph on $\mathcal{O}(2^{n/3})$ vertices in time $\mathcal{O}^*(2^{\omega n/3})$. As the auxiliary graph can be computed in time $\mathcal{O}^*(2^{2n/3})$, the total running time of the algorithm is $\mathcal{O}^*(2^{\omega n/3}) = \mathcal{O}(1.7315^n)$.

## 1.2.9   Partitioning Based Algorithms

Whereas the previous technique reduced an instance to an exponentially large instance of a problem solvable in polynomial time, this technique reduces an instance to an exponential number of simpler problems. Angelsmark et al. [AJ03, AT06, Ang05] designed several algorithms according to the slogan

> Solving an exponential number of small instances can be faster than solving a single large one.

For problems with domain size $d$, like $(d, l)$-CSP or $d$-COLORING, split the domain of each variable in different parts of given sizes in each possible way and solve all the corresponding subinstances using algorithms for smaller domains. Algorithms for $k$-COLORING, counting versions of $(d, 2)$-CSP and a variety of related problems have been obtained by this method; see the PhD thesis of Angelsmark [Ang05] for details.

## 1.2.10   Treewidth

Many $\mathcal{NP}$-hard graph problems can be solved in polynomial, or even linear time, when the input graph has constant treewidth.[7] Moreover, given a tree decomposition of width $\ell$ of a graph $G$, many treewidth based dynamic programming algorithms have a running time of the form $c^\ell \cdot n^{\mathcal{O}(1)}$ for some constant $c > 1$ [ABF+02]. As the treewidth of any planar graph is at most $\mathcal{O}(\sqrt{n})$, subexponential $2^{\mathcal{O}(\sqrt{n})}$ time algorithms are easily obtained for planar graphs. Similarly, bounds on the treewidth for sparse graphs lead directly to fast exponential time algorithms for sparse graphs. We refer to the survey by Fomin et al. [FGK05] for more details on direct implications of treewidth bounds. In Chapters 6, 7 and 8 we present several bounds on the treewidth of graphs and use these bounds in algorithms that combine branching and dynamic programming over tree decompositions.

## 1.2.11   Inclusion–Exclusion

The principle of Inclusion–Exclusion is a well known counting principle used to determine the cardinality of a union of overlapping sets. Namely if $V_1, V_2, \ldots, V_m$ are finite sets, then

$$\left| \bigcup_{i=1}^{m} V_i \right| = \sum_{i=1}^{m} |V_i| - \sum_{1 \le i < j \le m} |V_i \cap V_j| + \sum_{1 \le i < j < k \le m} |V_i \cap V_j \cap V_k| - \cdots + (-1)^{m-1} \left| \bigcap_{i=1}^{m} V_i \right|.$$

To our knowledge, the Inclusion–Exclusion principle was first used by Karp [Kar82] to design exponential time algorithms. An astonishing result using this principle was achieved by Björklund et al. [BHK09] for COLORING and other partitioning problems.

---

[7] The notion of treewidth is defined and discussed more extensively in Chapter 6. For the moment it is sufficient to know that it is a graph parameter measuring how tree-like a graph is.

Let us describe an $\mathcal{O}^*(2^n)$ algorithm for COLORING due to Björklund et al. [BHK09]. First, we prove that a graph $G = (V, E)$ is $k$-colorable if and only if $c_k(G) > 0$, where

$$c_k(G) := \sum_{X \subseteq V} (-1)^{|X|} s(X)^k$$

and $s(X)$ is the number of non–empty independent sets in $G$ not intersecting $X$. We will show that $c_k(G)$ is precisely the number of ways to cover $V$ with $k$ (possibly overlapping) non–empty independent sets. Note that $s(X)^k$ is the number of ways to choose, with repetition, $k$ non–empty independent sets not intersecting $X$. A set $C$ of $k$ non–empty independent sets covering $V$ is counted only in the term $(-1)^0 s(\emptyset)$, whereas a set $\bar{C}$ of $k$ non–empty independent sets not covering $V$ avoids some set of vertices $U$. Hence, $\bar{C}$ is counted once in every $s(W)$ for every $W \subseteq U$. As every non–empty set has as many even– as odd–sized subsets, the positive and negative counts of $\bar{C}$ sum up to 0. This shows that $G$ is $k$-colorable if and only if $c_k(G) > 0$.

The values $s(X)$ can then be computed by dynamic programming in time $\mathcal{O}^*(2^n)$. Let us instead compute $\bar{s}(X)$, which denotes the number of non–empty independent sets in $G[X]$. It is clear that $s(X) = \bar{s}(V \setminus X)$ for every $X \subseteq V$. By increasing cardinality of the sets $X$, compute $\bar{s}(X)$ using the formula

$$\bar{s}(X) := \begin{cases} 0, & \text{if } X = \emptyset, \text{ and} \\ \bar{s}(X \setminus v) + \bar{s}(X \setminus N[v]) + 1 \text{ for some } v \in X, & \text{otherwise.} \end{cases}$$

For the base case, it is clear that the number of non–empty independent sets in an empty graph is 0. Otherwise, each non–empty independent set counted in $\bar{s}(X)$ either does not contain $v$ and is counted in $\bar{s}(X \setminus v)$, or contains $v$ and is counted in $\bar{s}(X \setminus N[v]) + 1$ where the term $+1$ accounts for the singleton $\{v\}$.

Now, $c_k(G)$ can easily be computed. To obtain the least $k$ for which $c_k(G) > 0$, binary search may be used to solve COLORING in time $\mathcal{O}^*(2^n)$.

This technique has been further used and extended, for example to Subset Convolution via Möbius transform and inversion by Björklund et al. [BHKK07, BHKK08b, BHKK08c, BHKK08a] and to a kind of Inclusion–Exclusion branching by Nederlof, van Dijk, and van Rooij [Ned09, vRNvD09].

## 1.2.12 Parameterized Complexity and Parameter Bounded Subroutines

Parameterized Complexity [DF99, FG06, Nic06] is probably the area that is closest to exponential time algorithms. Here, one asks if given a problem and a parameter $k$, the problem can be solved in time $f(k) \cdot n^{\mathcal{O}(1)}$ where $f(\cdot)$ is an arbitrary computable function. One of the basic examples in Parameterized Complexity is the MINIMUM VERTEX COVER problem with $k$ being the cardinality of the vertex cover one is asked to find. The fastest (in terms of the function $f(k)$) known parameterized algorithm for this problem has running time $\mathcal{O}(1.2738^k + kn)$ [CKX06]. It is natural to use these algorithms as subroutines for exponential time algorithms in case a good bound on the parameter is known, as done for example in [Wah04, Wah07] for solving MINIMUM 3-HITTING SET and in [RSS05] for a variety of other problems.

On the negative side, for many problem/parameter combinations, it is unlikely that fixed parameter algorithms exist. When they introduced Parameterized Complexity in 1992, Downey and Fellows [DF92] defined the complexity classes

$$\mathcal{FPT} \subseteq \mathcal{W}[1] \subseteq \mathcal{W}[2] \subseteq \cdots \subseteq \mathcal{W}[P].$$

Problem/parameter combinations in $\mathcal{FPT}$ are fixed parameter tractable, that is, there exist algorithms with running times of the form $f(k) \cdot n^{\mathcal{O}(1)}$ for them. For problem/parameter combinations that are complete for one of the other classes, like MAXIMUM INDEPENDENT SET (where $k$ is the size of the independent set) which is complete for $\mathcal{W}[1]$ or MINIMUM DOMINATING SET (where $k$ is the size of the dominating set) which is complete for $\mathcal{W}[2]$, no algorithms with running times of the form $f(k) \cdot n^{\mathcal{O}(1)}$ are known, and it is strongly believed that none of these problems is in $\mathcal{FPT}$. Nevertheless, even algorithms with running time $\mathcal{O}(n^k)$ could be of interest for exponential time algorithms. For example, the notoriously hard BANDWIDTH problem, which is hard for every level of the $\mathcal{W}[\cdot]$ hierarchy even for trees [BFH94], not approximable in polynomial time within any constant [Ung98], and for which the fastest known exponential time algorithm has running time $\mathcal{O}^*(4.473^n)$ [CP09b], can be solved in time $\mathcal{O}^*(2^n)$ if the bandwidth is at most $n/\log n$ by an $\mathcal{O}(n^{k+1})$ algorithm of Saxe [Sax80].

### 1.2.13 Iterative Compression

Introduced in the area of Parameterized Complexity, this technique will be presented in detail in Chapter 9 and used to obtain faster algorithms for variants of the MINIMUM HITTING SET problem. For minimization problems, the basic idea is to build an iterative algorithm, and at each iteration, either compress the solution into a smallest one or prove that the current solution is optimal.

## 1.3 On Space and on Time

Several of the techniques presented in the previous section naturally lead to algorithms whose space complexity is close to their time complexity. A convenient and important theoretical classification is to distinguish between algorithms needing only polynomial space (in the worst case) and algorithms with exponential space complexity. Research goes in both directions here. For problems where fast exponential space algorithms are known, alternative methods are studied requiring only polynomial space, as in [GS87, FV08] for example. For problems where fast polynomial space algorithms are known, methods to trade space for time are applied to improve the running time bounds, as for example in [Rob86, FGK09b].

Whether or not exponential space algorithms are useful in practice is debatable. In one of his surveys on exponential time algorithms, Woeginger [Woe04] writes:

> Note that algorithms with exponential space complexities are absolutely useless

for real life applications.

On the other hand, after having implemented exponential space algorithms for BANDWIDTH, Cygan and Pilipczuk [CP08] write:

> It is worth mentioning that exponential space in our algorithms is no problem
> for practical implementations, since in every case space bound is less than square
> root of the time bound, thus space will not be a bottleneck in a real life applications,
> at least considering today's proportions of computing speed and fast memory size.

This shows that the community has not yet reached a consensus about what is practical regarding space consumption of the algorithms. Indeed, it seems that exponential time algorithms that require as much space as time, are impractical from the implementation point of view. But what about algorithms that require exponential space that is significantly (by a large enough exponential function) smaller than the running time of the algorithm? A good compromise seems to be achieved by techniques like Memorization or the techniques developed in Chapters 7 and 8 combining polynomial space branching and exponential space treewidth based algorithms. Here, the space complexity of the algorithm can be adjusted, by simply tuning a parameter of the algorithm, to the amount of space that one is willing to use, at the expense of a slightly higher running time.

Concerning the running time of exponential time algorithms, the question of what is a 'significant' improvement in the running time of an algorithm often arises. Consider for example (a) an improvement of a running time from $2.2^n$ to $2.1^n$, and (b) an improvement from $1.2^n$ to $1.1^n$. At a first glance, both improvements seem comparable. At a second look however, improvement (b) makes it now possible to solve instances that are $\log(1.2)/\log(1.1) \simeq 1.912$ times larger than before if both the $1.1^n$ and the $1.2^n$ algorithms are allowed to run for the same amount of time, whereas improvement (a) makes it now possible to solve instances that are only $\log(2.2)/\log(2.1) \simeq 1.062$ times larger. This is actually better seen if we use a $2^{c \cdot n}$ notation to express the running time of the algorithms and compare the constant in the exponent: $2.2^n = 2^{1.1376n}$, $2.1^n = 2^{1.0704n}$, $1.2^n = 2^{0.2631n}$, and $1.1^n = 2^{0.1376n}$.

Polynomial factors in the running time are hidden by the $\mathcal{O}^*$ notation. By rounding the base of the exponent, polynomial factors even disappear in the usual $\mathcal{O}$ notation; for example $\mathcal{O}(n^{100} \cdot 1.38765^n) = \mathcal{O}(1.3877^n)$. For practical implementations, however, care should be taken about the polynomial factors. Consider, for example, an algorithm $A$ with running time $T_A(n) = n \cdot 1.5^n$ and an algorithm $B$ with running time $T_B(n) = n^2 \cdot 1.46^n$. Algorithm $B$ seems preferable as, asymptotically, it is exponentially faster than Algorithm $A$. However, simple calculations show that for $n \leq 195$, Algorithm $A$ is faster; and for $n = 195$, both algorithms need to perform a number of operations which exceeds the number of attoseconds[8] since the Big Bang.

Table 1.1 gives an indication on the size of the instances that can be solved in a given amount of time, assuming that $2^{30}$ operations can be carried out per second, which roughly

---

[8]One attosecond is $10^{-18}$ seconds. The shortest time interval ever measured is about 100 attoseconds.

| Available time | 1 s | 1 min | 1 hour | 3 days | > 6 months |
|---|---|---|---|---|---|
| number of operations | $2^{30}$ | $2^{36}$ | $2^{42}$ | $2^{48}$ | $2^{54}$ |
| $n^5$ | 64 | 145 | 329 | 774 | 1756 |
| $n^{10}$ | 8 | 12 | 18 | 27 | 41 |
| $1.05^n$ | 426 | 510 | 594 | 681 | 765 |
| $1.1^n$ | 218 | 261 | 304 | 348 | 391 |
| $1.5^n$ | 51 | 61 | 71 | 82 | 92 |
| $2^n$ | 30 | 36 | 42 | 48 | 54 |
| $5^n$ | 12 | 15 | 18 | 20 | 23 |
| $n!$ | 12 | 14 | 15 | 17 | 18 |

Table 1.1: Maximum size of $n$ for different running times for a given amount of time under the assumption that $2^{30}$ operations can be performed in one second

corresponds to the computational power of an old Intel Pentium III processor at 500 MHz from 1999. Note that technology is not the predominant factor if we would like to exactly solve $\mathcal{NP}$-hard problems in practice, but algorithms are. Suppose there is an algorithm to solve some problem in time $2^n$ and the current implementation has an acceptable running time for instance sizes up to a constant $x$. To solve larger instances, we can wait for faster technology and solve instances of size $x + 1$ after a period of 18–24 months, according to Moore's law, or we can try to find faster algorithms; a $1.7548^n$ algorithm would solve the problem for instances up to size $1.23 \cdot x$. Moreover, for instances of size, say 50, the $1.7548^n$ algorithm performs 692 times faster than the $2^n$ algorithm.

It is also worth noticing that some of the techniques presented in Section 1.2, like dynamic programming, inherently lead to algorithms that have the same performance on every instance of a given size $n$, whereas other techniques, like branching, lead to algorithms that naturally perform much faster on most inputs than in a worst case scenario. Moreover, for most branching algorithms, the known lower bounds on their running time are far from the upper bound and it is very well possible that the worst case running time of most branching algorithms is overestimated.

## 1.4 Outline of the Book

The individual chapters of this book are organized as follows.

Chapters 2–5 mainly focus on branching algorithms. A general introduction to branching algorithms is given in Chapter 2. The main focus of that chapter is the running time analysis of branching algorithms, presented in a very general way, and providing a solid background for the following chapters. Chapter 3 presents an $\mathcal{O}(1.7548^n)$ algorithm for the FEEDBACK VERTEX SET problem, as well as upper bound of $1.8638^n$ and a lower bound of $1.5926^n$ on the number of minimal feedback vertex sets in a graph. The upper bound on the number of feedback vertex sets is obtained via the same kind of techniques that are used to upper bound the running time of the FEEDBACK VERTEX SET algorithm, and the lower bound is achieved by an explicit construction of an infinite family of graphs with a large number minimal feedback vertex sets. In Chapter 4, the focus is on bicliques in graphs. We transform different results for independent sets to results concerning bicliques and derive a $n \cdot 3^{n/3}$ upper bound on the number of maximal bicliques in a graph on $n$ vertices, derive algorithms that can find, count and enumerate maximal or maximum bicliques in the same time bound (up to polynomial factors) as the corresponding algorithms for independent sets. For lack of an existing algorithm able to count all maximal independent sets faster than enumerating all of them, we also provide the first non trivial algorithm counting all maximal independent sets in a graph. It has running time $\mathcal{O}(1.3642^n)$. The last example of a branching algorithm is presented in Chapter 5. Here we present the currently fastest polynomial space algorithm for MAX 2-SAT, MAX 2-CSP and mixed instances of these two problems. A very rigorous analysis allows us to use both MAX 2-SAT and MAX 2-CSP specific simplification rules and to parameterize the running time by the fraction of general integer–weighted CSP clauses versus simple unit–weighted conjunctions and disjunctions to obtain the currently fastest polynomial space algorithm for MAX 2-SAT, MAX 2-CSP, and everything in between.

In Chapters 6–8, we present bounds for the treewidth of graphs and two Win–Win strategies combining branching and tree decomposition based algorithms. Treewidth bounds that are needed in the two following chapters are proved in Chapter 6. In particular, we give upper bounds for the treewidth in terms of the number of vertices of sparse graphs and in terms of the maximum degree of circle graphs. In Chapter 7 we derive faster algorithms for the MINIMUM DOMINATING SET problem for graph classes where one can find tree decompositions whose widths do not exceed the maximum degree of the graph by more than a constant multiplicative factor. A general framework based on the enumeration of independent sets and tree decomposition based algorithms for sparse graphs is presented in Chapter 8 and applied to different problems.

Showing how to carry over to exponential time algorithms a technique that is prominent in the area of Parameterized Complexity, Chapter 9 presents exponential time algorithms based on iterative compression for MINIMUM HITTING SET–like problems.

Finally, Chapter 10 concludes with a short summary, some open problems and possible further research directions.

# Branching Algorithms

> Nothing is particularly hard if you divide it into small jobs.

*Henry Ford*

In this chapter, we present branching algorithms and various ways to analyze their running times. Unlike for other techniques to design exponential time algorithms (or polynomial time algorithms), like dynamic programming, it is far less obvious for branching algorithms how to obtain worst case running time bounds that are (close to) tight. The running time analysis is a very important and non trivial factor in the design of branching algorithms; the design and the analysis of branching algorithms usually influence each other strongly and they go hand in hand. Also, branching algorithms usually perform faster on real life data and randomized instances than the (upper bound of the) worst case running time derived by the analysis: it is not uncommon that competitive SAT solvers in competitions like SAT Race and SAT Competition solve instances with thousands of variables, although no known algorithm for SAT has a proved (worst case) time complexity of $\mathcal{O}(1.9999^n)$.

Branching algorithms are recursive algorithms that solve a problem instance by reducing it to "smaller" instances, solving these recursively, and combining the solutions of the subinstances to a solution for the original instance.

In the literature they appear with various names, for example Branch-and-Reduce, Branch-and-Bound, Branch-and-Search, Branching, Pruning the Search Tree, Backtracking, DPLL, or Splitting algorithms.

Typically, these algorithms

1. select a local configuration of the problem instance (selection),

2. determine the possible values this local configuration can take (inspection),

3. recursively solve subinstances based on these values (recursion), and

4. compute an optimal solution of the instance based on the optimal solutions of the subinstances (combination).

---

Algorithm **mis**($G$)
**Input**   : A graph $G = (V, E)$.
**Output**: The size of a maximum independent set of $G$.

1  **if** $\Delta(G) \leq 2$ **then**                                        // $G$ has maximum degree at most 2
2      **return** the size of a maximum independent set of $G$ in polynomial time
3  **else if** $\exists v \in V : d(v) = 1$ **then**                        // $v$ has degree 1
4      **return** $1 + \mathbf{mis}(G \backslash N[v])$
5  **else if** $G$ is not connected **then**
6      Let $G_1$ be a connected component of $G$
7      **return** $\mathbf{mis}(G_1) + \mathbf{mis}(G \setminus V(G_1))$
8  **else**
9      Select $v \in V$ such that $d(v) = \Delta(G)$                      // $v$ has maximum degree
10     **return** $\max\left(1 + \mathbf{mis}(G \setminus N[v]), \mathbf{mis}(G \setminus v)\right)$

Figure 2.1: Algorithm **mis**($G$), computing the size of a maximum independent set of any input graph $G$

---

We call *reduction* a transformation (selection, inspection and the creation of the subinstances for the recursion) of the initial instance into one or more subinstances. We also call *simplification* a reduction to one subinstance, and *branching* or *splitting* a reduction to more than one subinstance.

Usually, the reduction and the combination steps take polynomial time and the reduction creates a constant number of subinstances (for exceptions, see for example [Ang05]). Polynomial time procedures to solve a problem for "simple" instances are viewed here as simplification rules, reducing the instance to the empty instance.

Let us illustrate branching algorithms by a simple algorithm for MAXIMUM INDEPENDENT SET. Consider Algorithm **mis** on this page. It contains two simplification rules. The first one (lines 1–2) solves MAXIMUM INDEPENDENT SET for graphs of maximum degree 2 in polynomial time. Clearly, for a collection of paths and cycles, the size of a maximum independent set can be computed in polynomial time: a maximum independent set of $P_n$ has size $\lceil n/2 \rceil$ and a maximum independent set of $C_n$ has size $\lfloor n/2 \rfloor$. The second simplification rule (lines 3–4) always includes vertices of degree 1 in the considered independent set. Its correction is based on the following observation.

**Observation 2.1.** *For a vertex $v$ of degree 1, there exists a maximum independent set containing $v$.*

*Proof.* Suppose not, then all maximum independent sets of $G$ contain $v$'s neighbor $u$. But then we can select one maximum independent set, replace $u$ by $v$ in this independent set, resulting in an independent set of the same size and containing $v$ — a contradiction.                      □

By the argument used in the proof of Observation 2.1, the algorithm is not guaranteed to go through all maximum independent sets of $G$, but is guaranteed to find at least one of them.

The first branching rule (lines 5–7) is invoked when $G$ has at least two connected components. Clearly, the size of a maximum independent set of a graph is the sum of the sizes of the maximum independent sets of its connected components. If $V(G_1)$ (or $V \setminus V(G_1)$) has constant size, this branching rule may actually be viewed as a simplification rule, as $G_1$ (or $G \setminus V(G_1)$) is dealt with in constant time.

The second branching rule (lines 8–10) of the algorithm selects a vertex $v$ of maximum degree; this vertex corresponds to the local configuration of the problem instance that is selected. Including or excluding this vertex from the current independent set are the values that this local configuration can take. The subinstances that are solved recursively are $G \setminus N[v]$ — including the vertex $v$ in the independent set, which prevents all its neighbors to be included — and $G \setminus v$ — excluding $v$ from the independent set. Finally, the computation of the maximum in the last line of the algorithm corresponds to the combination step.

## 2.1 Simple Analysis

To derive upper bounds for the running time of a branching algorithm, let us describe its behavior by a model which consists of a set of univariate constraints.

**Definition 2.2.** Given an algorithm $A$ and an instance $I$, $T_A(I)$ denotes the running time of $A$ on instance $I$.

**Lemma 2.3.** *Let $A$ be an algorithm for a problem $P$, and $\alpha > 0$, $c \geq 0$ be constants such that for any input instance $I$, $A$ reduces $I$ to instances $I_1, \ldots, I_k$, solves these recursively, and combines their solutions to solve $I$, using time $\mathcal{O}(|I|^c)$ for the reduction and combination steps (but not the recursive solves),*

$$(\forall i : 1 \leq i \leq k) \quad |I_i| \leq |I| - 1, \text{ and} \tag{2.1}$$

$$\sum_{i=1}^{k} 2^{\alpha \cdot |I_i|} \leq 2^{\alpha \cdot |I|}. \tag{2.2}$$

*Then $A$ solves any instance $I$ in time $\mathcal{O}(|I|^{c+1}) 2^{\alpha \cdot |I|}$.*

*Proof.* The result follows easily by induction on $|I|$. Without loss of generality, we may replace the hypotheses' $\mathcal{O}$ statements with simple inequalities (substitute a sufficiently large leading constant, which then appears everywhere and has no relevance), and likewise for the base case assume that the algorithm returns the solution to an empty instance in time $1 \leq |I|^{c+1} 2^{\alpha \cdot |I|}$.

Suppose the lemma holds for all instances of size at most $|I| - 1 \geq 0$, then

$$T_A(I) \leq |I|^c + \sum_{i=1}^{k} T_A(I_i) \qquad \text{(by definition)}$$

$$\leq |I|^c + \sum |I_i|^{c+1} 2^{\alpha \cdot |I_i|} \qquad \text{(by the inductive hypothesis)}$$

$$\leq |I|^c + (|I| - 1)^{c+1} \sum 2^{\alpha \cdot |I_i|} \qquad \text{(by (2.1))}$$

$$\leq |I|^c + (|I| - 1)^{c+1} 2^{\alpha \cdot |I|}$$  (by (2.2))

$$\leq |I|^{c+1} 2^{\alpha \cdot |I|}.$$

The final equality uses that $\alpha \cdot |I| > 0$ and holds for any $c \geq 0$.                       □

Let us use this lemma to derive a vertex-exponential upper bound of the running time of Algorithm **mis**, executed on a graph $G = (V, E)$ on $n$ vertices. For this purpose we set $|G| := |V| = n$. We may at all times assume that $n$ is not a constant, otherwise the algorithm takes constant time.

Determining if $G$ has maximum degree 2 can clearly be done in time $\mathcal{O}(n)$. By a simple depth–first–search, and using a pointer to the first unexplored vertex, the size of a maximum independent set for graphs of maximum degree 2 can also be computed in time $\mathcal{O}(n)$. Checking if $G$ has more than one connected component can be done in time $\mathcal{O}(n + m) = \mathcal{O}(n^2)$. Finding a vertex of maximum degree and the creation of the two subinstances takes time $\mathcal{O}(n + m) = \mathcal{O}(n^2)$. Addition and the computation of the maximum of two numbers takes time $\mathcal{O}(1)$.

For the first branching rule, we obtain a set of constraints for each possible size $s$ of $V(G_1)$:

$$(\forall s : 1 \leq s \leq n - 1) \quad 2^{\alpha \cdot s} + 2^{\alpha \cdot (n-s)} \leq 2^{\alpha \cdot n}.$$  (2.3)

By convexity of the function $2^x$, these constraints are always satisfied, irrespective of the value of $\alpha > 0$ and can thus be ignored. Here we suppose that $n$ is not a constant, otherwise the algorithm takes only constant time.

For the second branching rule, we obtain a constraint for each vertex degree $d \geq 3$:

$$(\forall d : 3 \leq d \leq n - 1) \quad 2^{\alpha \cdot (n-1)} + 2^{\alpha \cdot (n-1-d)} \leq 2^{\alpha n}.$$  (2.4)

Dividing all these terms by $2^{\alpha n}$, the constraints become

$$2^{-\alpha} + 2^{\alpha \cdot (-1-d)} \leq 1.$$  (2.5)

Then, by standard techniques [Kul99], the minimum $\alpha$ satisfying all these constraints is obtained by setting $x := 2^\alpha$, computing the unique positive real root of each of the characteristic polynomials

$$c_d(x) := x^{-1} + x^{-1-d} - 1,$$

by Newton's method, for example, and taking the maximum of these roots. Alternatively, one could also solve a mathematical program minimizing $\alpha$ subject to the constraints in (2.5) (the constraint for $d = 3$ is sufficient as all other constraints are weaker). The maximum of these roots (see Table 2.1) is obtained for $d = 3$ and its value is $1.380277\ldots \approx 2^{0.464958\ldots}$.

Applying Lemma 2.3 with $c = 2$ and $\alpha = 0.464959$, we find that the running time of Algorithm **mis** is upper bounded by $\mathcal{O}(n^3) \cdot 2^{0.464959 \cdot n} = \mathcal{O}(2^{0.46496 \cdot n}) = \mathcal{O}(1.3803^n)$. Here the $\mathcal{O}$ notation permits to exclude the polynomial factor by rounding the exponential factor.

| $d$ | $x$ | $\alpha$ |
|---|---|---|
| 3 | 1.3803 | 0.4650 |
| 4 | 1.3248 | 0.4057 |
| 5 | 1.2852 | 0.3620 |
| 6 | 1.2555 | 0.3282 |
| 7 | 1.2321 | 0.3011 |

Table 2.1: Positive real roots of $c_d(x)$

## 2.2 Lower Bounds on the Running Time of an Algorithm

One peculiarity of branching algorithms is that it is usually not possible, by the currently available running time analysis techniques, to match the derived upper bound of the running time by a problem instance for which the algorithm really takes this much time to compute a solution. Exceptions are brute-force branching algorithms that go through all the search space or algorithms enumerating objects for which tight bounds on their number are known, like the enumeration of maximal independent sets [MM65] or maximal induced bicliques [GKL08].

Lower bounds on the running time of a specific algorithm are helpful as they might give indications which instances are hard to solve by the algorithm, an information that might suggest attempts to improve the algorithm. Moreover, the design (or its attempt) of lower bound instances might give indications on which "bad" structures do not exist in an instance unless some other "good" structures arise during the execution of the algorithm, which might hint at the possibility of a better running time analysis. Finally, it is desirable to sandwich the true worst case running time of the algorithm between an upper and a lower bound to obtain more knowledge on it as a part of the analysis of the algorithm.

Lower bounds are usually obtained by describing an infinite family of graphs for which the behavior of the algorithm is "cyclic", that is it branches on a finite number of structures in the instance in a periodic way.

To derive a lower bound of the running time of Algorithm **mis**, consider the graph $G = P_n^2$, depicted in Figure 2.2 — the second power of a path on $n$ vertices, obtained from a path $P_n$ on $n$ vertices by adding edges between every two vertices at distance at most two in $P_n$. Suppose $n \geq 9$, then none of the simplification rules applies to $G$. The algorithm selects some vertex of degree 4; here, we — the designers of the lower bound — have the freedom to make the algorithm choose a specific vertex of degree 4, as it does not itself give any preference. Suppose therefore, that it selects $v_3$ to branch on. It creates two subproblems:

- $G \setminus N[v_3]$, a graph isomorphic to $P_{n-5}^2$, and

- $G \setminus v_3$, a graph isomorphic to a $P_2$ connected with one edge to the first vertex of a $P_{n-3}^2$. In the first recursive step of **mis**, the reduction rule on vertices of degree at most 1 includes $v_1$ in the independent set and recurses on the graph isomorphic to $P_{n-3}^2$.

Now, on each of these subproblems of sizes $n - 3$ and $n - 5$, the behavior of the algorithm is again the same as for the original instance of size $n$. Therefore, the running time of the

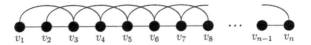

Figure 2.2: Graph $P_n^2$ used to lower bound the running time of Algorithm **mis**

---

**if** $\exists u, v \in V : N[u] \subseteq N[v]$ **then**
$\quad \lfloor$ **return mis**$(G \setminus v)$

Figure 2.3: Additional simplification rule for Algorithm **mis**

---

algorithm can be lower bounded by $\Omega(x^n)$ where $x$ is the positive root of

$$x^{-3} + x^{-5} - 1,$$

which is $1.193859\ldots \approx 2^{0.255632\cdots}$. This gives a lower bound on the running time of Algorithm **mis** of $\Omega(1.1938^n)$.

Let us take a closer look at the decisions made by the algorithm with a $P_n^2$ as input. The size of the independent set increases by 1 when including $v_3$ and also by 1 when excluding $v_3$ and including $v_1$ by a simplification rule. But the instance obtained by the second recursive call is, after the application of the simplification rule, a subgraph of the instance of the first recursive call. Therefore, the second recursive call always leads to a solution that is at least as good as the one obtained in the first recursive call. This is a special case of a set of local configurations where there exist two vertices $u, v$ such that $N[u] \subseteq N[v]$. In such a situation, the algorithm can just exclude $v$ from being in the maximum independent set it computes: consider a maximum independent set $I_v$ of $G$ containing $v$, then $I_v \setminus \{v\} \cup \{u\}$ is a maximum independent set of $G$ not containing $v$.

Thus, we could enhance the algorithm by adding the corresponding simplification rule; see Figure 2.3 [1].

## 2.3  Measure Based Analysis

One drawback of the analysis presented in Section 2.1 is that, when reducing the problem instance to several subinstances, many structural changes in the instance are not accounted for by a simple analysis in just the instance size. Therefore, let us in this section use a potential–function method akin to the measures used by Kullmann [Kul99], the quasiconvex analysis of Eppstein [Epp06], the "Measure and Conquer" approach of Fomin et al. [FGK09b], the (dual to the) linear programming approach of Scott and Sorkin [SS07], and much older potential–function analyses in mathematics and physics.

For Algorithm **mis**, for example, the simple analysis does not take into account the decrease

---

[1]The simplification rule for vertices of degree 1 becomes obsolete by adding this rule as vertices of degree 1 always fall under the scope of the new rule

of the degrees of the neighbors when deleting a vertex from the graph. Taking this decrease of the degrees into account is particularly useful when deleting a vertex adjacent to vertices of degree 2: their degree decreases to 1 and they (as well as their neighbors) are removed by a simplification rule.

Therefore, let us in a first step model the worst case running time of the algorithm by a set of multivariate constraints, where the variables correspond to the structures whose changes we would like to take into account. These structures may depend as well on the input instance the algorithm is currently considering, as on the output it is currently generating. Examples of parameters that the analysis may rely on are the number of vertices/variables of certain degrees, the number of triangles in the input graph, the size of the current solution, (an estimation of) the treewidth of the input graph, the connectivity of the current solution, and possibly many others.

For notational convenience, let us make the multivariate constraints depend solely on the problem instance, not on both the problem instance and the currently computed solution. There is no loss of generality as the current solution can always be handed to the algorithm as a part of the input.

For Algorithm **mis**, let us model the worst case running time by $T(n_1, n_2, \ldots)$, abbreviated as $T\left(\{n_i\}_{i \geq 1}\right)$ where $n_i$ denotes the number of vertices of degree $i$ in $G$.

For the analysis of the second branching rule, let us take into account the decrease of the degrees of the neighbors of $v$ when deleting $v$ and the decrease by 1 of the degree of one vertex in $N^2(v)$ when deleting $N[v]$. [2] Let

- $d \geq 3$ be the degree of $v$, that is the maximum degree of $G$,

- $p_i, 2 \leq i \leq d$ such that $\sum_{i=2}^{d} p_i = d$ be the number of neighbors of $v$ of degree $i$, and

- $k$ such that $2 \leq k \leq d$ be the degree of a vertex in $N^2(v)$.

In the branch where $v$ is deleted, the number of vertices of degree $i$

- decreases by 1 if $d = i$ ($v$ is removed),

- decreases by $p_i$ (the neighbors of $v$ of degree $i$ have their degrees reduced), and

- increases by $p_{i+1}$ (the neighbors of $v$ of degree $i + 1$ have their degree reduced to $i$).

In the branch where $N[v]$ is deleted, the number of vertices of degree $i$

- decreases by 1 if $d = i$ ($v$ is removed),

- decreases by $p_i$ (the neighbors of $v$ of degree $i$ are removed),

- decreases by 1 if $k = i$ (a vertex in $N^2(v)$ of degree $i$ has its degree reduced), and

---

[2] We assume for a moment that $v$ has at least one vertex at distance 2. It will become clear that this assumption is reasonable when we restrict the analysis to graphs of maximum degree 5. For a vertex $v$ of degree at most 5, $|N^2(v)| \geq 1$ if $G$ has at least 7 vertices (for constant size instances our algorithm runs in constant time), because $G$ is connected.

- increases by 1 if $k = i + 1$ (a vertex in $N^2(v)$ of degree $i + 1$ has its degree reduced to $i$).

Thus, we obtain the following recurrence where the maximum ranges over all $d \geq 3$, all $p_i, 2 \leq i \leq d$ such that $\sum_{i=2}^{d} p_i = d$ and all $k$ such that $2 \leq k \leq d$:

$$T\left(\{n_i\}_{i \geq 1}\right) = \max_{d, p_2, \ldots, p_d, k} \begin{cases} T(\{n_i - p_i + p_{i+1} - \mathsf{K}_\delta(d = i)\}_{i \geq 1}) \\ + T(\{n_i - p_i - \mathsf{K}_\delta(d = i) - \mathsf{K}_\delta(k = i) + \mathsf{K}_\delta(k = i + 1)\}_{i \geq 1}) \end{cases} \tag{2.6}$$

Here, $\mathsf{K}_\delta(\cdot)$ is the logical Kronecker delta [CB94], returning 1 if its argument is true and 0 otherwise.

*Remark* 1. This model of the worst case running time of the algorithm makes the assumption that it is always beneficial to decrease the degree of a vertex. When $N[v]$ is deleted, many more vertices in $N^2(v)$ could have their degree reduced (also by more than 1). Such assumptions are often necessary to limit the number of cases that need to be considered in the analysis.

In order to make the number of terms in recurrence (2.6) finite, let us restrict the running time analysis to graphs of maximum degree 5 in this section. In Section 2.7, we will combine an analysis for graphs of maximum degree 5 with the simple analysis of the previous section to derive an overall running time for Algorithm **mis** for graphs of arbitrary degrees.

Based on the multivariate recurrence (2.6) for $3 \leq d \leq 5$, we would now like to compute an upper bound of the algorithm's running time. Eppstein [Epp06] transforms the models based on multivariate recurrences into models based on weighted univariate linear recurrences and shows that there exists a set of weights for which the solution of one model is within a polynomial factor of the solution of the other model.

**Definition 2.4.** A *measure* $\mu$ for a problem $P$ is a function from the set of all instances for $P$ to the set of non negative reals.

To analyze Algorithm **mis**, let us use the following measure of a graph $G$ of maximum degree 5, which is obtained by associating a weight to each parameter $\{n_i\}_{1 \leq i \leq 5}$:

$$\mu(G) := \sum_{i=1}^{5} w_i n_i, \tag{2.7}$$

where $w_i \in \mathbb{R}^+$ for $i \in \{1, \ldots, 5\}$ are the weights associated with vertices of different degrees.

With this measure and the tightness result of Eppstein, we may transform recurrence (2.6) into a univariate recurrence. For convenience, let

$$(\forall d : 2 \leq d \leq 5) \quad h_d := \min_{2 \leq i \leq d} \{w_i - w_{i-1}\}, \tag{2.8}$$

denote the minimum decrease of $\mu(G)$ when reducing by 1 the degree of a vertex of degree at least 2 and at most $d$.

By the result of Eppstein there exist weights $w_i$ such that a solution to (2.6) corresponds to a solution to the following recurrence (for an optimal assignment of the weights), where the maximum ranges over all $d, 3 \leq d \leq 5$, and all $p_i, 2 \leq i \leq d$, such that $\sum_{i=2}^{d} p_i = d$,

$$T\left(\mu(G)\right) = \max_{d, p_2, \ldots, p_d, k} \begin{cases} T\left(\mu(G) - w_d - \sum_{i=2}^{d} p_i \cdot (w_i - w_{i-1})\right) \\ +T\left(\mu(G) - w_d - \sum_{i=2}^{d} p_i \cdot w_i - h_d\right). \end{cases} \tag{2.9}$$

The solution to (2.9) clearly satisfies the following constraints for all $d, 3 \leq d \leq 5$, and all $p_i, 2 \leq i \leq d$, such that $\sum_{i=2}^{d} p_i = d$:

$$T\left(\mu(G)\right) \geq T\left(\mu(G) - w_d - \sum_{i=2}^{d} p_i \cdot (w_i - w_{i-1})\right)$$
$$+ T\left(\mu(G) - w_d - \sum_{i=2}^{d} p_i \cdot w_i - h_d\right). \tag{2.10}$$

In order to upper bound the running time of algorithms based on a more involved measure of the size of an instance and constraints like the ones in (2.10), let us prove a lemma analogous to Lemma 2.3 on page 31 in Section 2.1.

**Lemma 2.5.** *Let $A$ be an algorithm for a problem $P$, $c \geq 0$ be a constant, and $\mu(\cdot), \eta(\cdot)$ be measures for the instances of $P$, such that for any input instance $I$, $A$ reduces $I$ to instances $I_1, \ldots, I_k$, solves these recursively, and combines their solutions to solve $I$, using time $\mathcal{O}(\eta(I)^c)$ for the reduction and combination steps (but not the recursive solves),*

$$(\forall i) \quad \eta(I_i) \leq \eta(I) - 1, \text{ and} \tag{2.11}$$

$$\sum_{i=1}^{k} 2^{\mu(I_i)} \leq 2^{\mu(I)}. \tag{2.12}$$

*Then $A$ solves any instance $I$ in time $\mathcal{O}(\eta(I)^{c+1})2^{\mu(I)}$.*

*Proof.* The result follows by induction on $\eta(I)$. As in the proof of Lemma 2.3, we may replace the hypotheses' $\mathcal{O}$ statements with simple inequalities. For the base case assume that the algorithm returns the solution to an empty instance in time $1 \leq \eta(I)^{c+1}2^{\mu(I)}$. Suppose the lemma holds for all instances $I'$ with $\eta(I') \leq \eta(I) - 1 \geq 0$, then

$$\begin{aligned}
T_A(I) &\leq \eta(I)^c + \sum_{i=1}^{k} T_A(I_i) && \text{(by definition)} \\
&\leq \eta(I)^c + \sum \eta(I_i)^{c+1}2^{\mu(I_i)} && \text{(by the inductive hypothesis)} \\
&\leq \eta(I)^c + (\eta(I) - 1)^{c+1}\sum 2^{\mu(I_i)} && \text{(by (2.11))} \\
&\leq \eta(I)^c + (\eta(I) - 1)^{c+1}2^{\mu(I)} && \text{(by (2.12))} \\
&\leq \eta(I)^{c+1}2^{\mu(I)}. && (c \geq 0 \text{ and } \mu(\cdot) \geq 0).
\end{aligned}$$

Thus the lemma follows. □

*Remark* 2. The measure $\eta(\cdot)$ corresponds often to the size of the input, but we will see in Chapter 3 an analysis for which $\eta(\cdot)$ is different. It is used to bound the depth of the recursion,

| $i$ | $w_i$ | $h_i$ |
|---|---|---|
| 0 | 0 | 0 |
| 1 | 0 | 0 |
| 2 | 0.25 | 0.25 |
| 3 | 0.35 | 0.10 |
| 4 | 0.38 | 0.03 |
| 5 | 0.40 | 0.02 |

Table 2.2: An assignment of the weights for the measure $\mu(G) = \sum_{i=1}^{5} w_i n_i$ for the analysis of Algorithm **mis**

whereas $\mu(\cdot)$ is used to bound the number of terminal recursive calls.

Slightly rephrasing (2.10) to fit into the framework of Lemma 2.5, we obtain the following set of constraints. For each $d, 3 \leq d \leq 5$, and all $p_i, 2 \leq i \leq d$, such that $\sum_{i=2}^{d} p_i = d$,

$$2^{\mu(G)} \geq 2^{\mu(G) - w_d - \sum_{i=2}^{d} p_i \cdot (w_i - w_{i-1})} + 2^{\mu(G) - w_d - \sum_{i=2}^{d} p_i \cdot w_i - h_d}. \tag{2.13}$$

Dividing by $2^{\mu(G)}$, we obtain

$$1 \geq 2^{-w_d - \sum_{i=2}^{d} p_i \cdot (w_i - w_{i-1})} + 2^{-w_d - \sum_{i=2}^{d} p_i \cdot w_i - h_d}. \tag{2.14}$$

With the values in Table 2.2 for $w_i$, all these constraints are satisfied. With these weights, $\mu(G) \leq 2n/5$ for any graph of maximum degree 5 on $n$ vertices. Taking $c = 2$ and $\eta(G) = n$, Lemma 2.5 implies that Algorithm **mis** has running time $\mathcal{O}(n^3)2^{2n/5} = \mathcal{O}(1.3196^n)$ on graphs of maximum degree at most 5.

Thus, we were able to improve the analysis of Algorithm **mis** by a different measure of graphs. However, the weights in Table 2.2 are not optimal for this model of the running time of Algorithm **mis** and we will discuss in the next section how to optimize the weights.

## 2.4   Optimizing the Measure

In the literature, mainly two techniques have been used to optimize the weights for measures employed for upper bounding the worst case running time of branching algorithms. These are a sort of random local search [FGK05, FGK09b] and quasiconvex programming [Epp06].

In this book, we will use convex programming to optimize the weights of the considered measures. As affine multivariate functions are convex, and the function $2^x$ is convex and non decreasing, and the composition $g \circ f$ of a convex, non decreasing function $g$ and a convex function $f$ is convex, and summing convex functions preserves convexity, all the constraints in Lemma 2.5 are convex if the measure $\mu$ is a linear function in the weights.

In order to compute the optimal weights for the improved analysis of Algorithm **mis**, let us use a standard trick in linear programming to make the conditions (2.8) on $h_d$ linear:

$$(\forall i, d : 2 \leq i \leq d) \quad h_d \leq w_i - w_{i-1}. \tag{2.15}$$

```
# Introductory example: Maximum Independent Set

# maximum degree
param maxd integer >= 3;
# all possible degrees
set DEGREES := 0..maxd;
# weight for vertices according to their degrees
var W {DEGREES} >= 0;
# weight for degree reductions from degree exactly i
var g {DEGREES} >= 0;
# weight for degree reductions from degree at most i
var h {DEGREES} >= 0;
# maximum weight of W[d]
var Wmax;

# minimize the maximum weight
minimize Obj: Wmax;

# the max weight is at least the weight for vertices of degree d
subject to MaxWeight {d in DEGREES}:
  Wmax >= W[d];

# constraints for the values of g[]
subject to gNotation {d in DEGREES : 2 <= d}:
  g[d] <= W[d]-W[d-1];

# constraints for the values of h[]
subject to hNotation {d in DEGREES, i in DEGREES : 2 <= i <= d}:
  h[d] <= W[i]-W[i-1];

# constraints for max degree 3
subject to Deg3 {p2 in 0..3, p3 in 0..3 : p2+p3=3}:
  2^(-W[3] - p2*g[2] - p3*g[3])
+ 2^(-W[3] - p2*W[2] - p3*W[3] - h[3]) <=1;

# constraints for max degree 4
subject to Deg4 {p2 in 0..4, p3 in 0..4, p4 in 0..4 : p2+p3+p4=4}:
  2^(-W[4] - p2*g[2] - p3*g[3] - p4*g[4])
+ 2^(-W[4] - p2*W[2] - p3*W[3] - p4*W[4] - h[4]) <=1;

# constraints for max degree 5
subject to Deg5 {p2 in 0..5, p3 in 0..5, p4 in 0..5, p5 in 0..5 : p2+p3+p4+p5=5}:
  2^(-W[5] - p2*g[2] - p3*g[3] - p4*g[4] - p5*g[5])
+ 2^(-W[5] - p2*W[2] - p3*W[3] - p4*W[4] - p5*W[5] - h[5]) <=1;
```

Figure 2.4: Mathematical program in AMPL modeling the constraints for the analysis of Algorithm **mis**

| $i$ | $w_i$ | $h_i$ |
|---|---|---|
| 0 | 0 | 0 |
| 1 | 0 | 0 |
| 2 | 0.206018 | 0.206018 |
| 3 | 0.324109 | 0.118091 |
| 4 | 0.356007 | 0.031898 |
| 5 | 0.358044 | 0.002037 |

Table 2.3: An optimal assignment of the weights for the measure $\mu(G) = \sum_{i=1}^{5} w_i n_i$ for the analysis of Algorithm **mis**

Minimizing the maximum $w_i, 0 \le i \le 5$ can thus be done by solving a convex program, whose implementation in AMPL (A Mathematical Programming Language) [FGK03] is depicted in Figure 2.4.

Using mathematical programming solvers such as IPOPT (part of the free, open-source code repository at `www.coin-or.org`) and MINOS (a commercial solver), the mathematical programs can be solved to optimality very fast (less than a second on a typical nowadays laptop for the program in Figure 2.4 with 95 constraints).

The mathematical program provides the optimal values for $w_i$, see Table 2.3. With these weights, $\mu(G) \le 0.358044 \cdot n$ for any graph of maximum degree 5 on $n$ vertices. Taking $c = 2$ and $\eta(G) = n$, Lemma 2.5 implies that Algorithm **mis** has running time $\mathcal{O}(n^3)2^{0.358044 \cdot n} = \mathcal{O}(1.2817^n)$.

The tight constraints for the second branching rule of the algorithm is inequality (2.14) with parameters

- $d = 3, p_2 = 3, p_3 = 0$,

- $d = 3, p_2 = 0, p_3 = 3$,

- $d = 4, p_2 = 0, p_3 = 0, p_4 = 4$, and

- $d = 5, p_2 = 0, p_3 = 0, p_4 = 0, p_5 = 5$.

Improving at least one of the tight constraints usually leads to a better upper bound for the worst case running time of the algorithm (except if there is another equivalent constraint).

## 2.5   Search Trees

The execution of a branching algorithm on a particular input instance can naturally be depicted as a *search tree* or *recursion tree*: a rooted tree where the root is associated to the input instance and for every node whose associated instance $I$ is reduced to the subinstances $I_1, I_2, \ldots, I_k$, it has children associated to these subinstances; see Figure 2.5. Often the nodes of the search tree are labeled with the measure of the corresponding instance.

For instance, Figure 2.6 shows a part of the search tree corresponding to the execution of Algorithm **mis** on the graph $P_n^2$, depicted in Figure 2.2 on page 34.

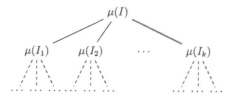

Figure 2.5: Illustration of a search tree

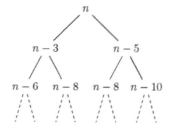

Figure 2.6: First levels of the search tree for the execution of Algorithm **mis** on the instance $P_n^2$

With the assumptions in Lemma 2.5 on page 37, the running time of an algorithm $A$ for a particular input instance $I$ is proportional — up to a polynomial factor — to the number of leaves in the corresponding search tree. To upper bound the running time of an algorithm, one could therefore bound the number of leaves in any search tree corresponding to the algorithm.

## 2.6 Branching Numbers and their Properties

Let us introduce a convenient and short way to specify the most common constraints that we use. Given a constraint of the form

$$2^{\mu(I)-a_1} + \cdots + 2^{\mu(I)-a_k} \leq 2^{\mu(I)}, \tag{2.16}$$

we define its *branching number* to be

$$2^{-a_1} + \cdots + 2^{-a_k}, \tag{2.17}$$

and denote it by

$$(a_1, \ldots, a_k). \tag{2.18}$$

Clearly, any constraint with branching number at most 1 is satisfied. The following two properties of branching numbers are useful to eliminate unnecessary constraints.

**Dominance** For any $a_i, b_i$ such that $a_i \geq b_i$ for all $i$, $1 \leq i \leq k$,

$$(a_1, \ldots, a_k) \leq (b_1, \ldots, b_k), \tag{2.19}$$

as $2^{-a_i} \leq 2^{-b_i}$ for all $i, 1 \leq i \leq k$. We say in this case that the branching number $(a_1, \ldots, a_k)$ is *dominated* by the branching number $(b_1, \ldots, b_k)$. In particular, this implies that for any $a, b > 0$,

$$\text{either} \quad (a, a) \leq (a, b) \quad \text{or} \quad (b, b) \leq (a, b). \tag{2.20}$$

**Balance** If $0 < a \leq b$, then for any $\varepsilon$ such that $0 \leq \varepsilon \leq a$,

$$(a, b) \leq (a - \varepsilon, b + \varepsilon) \tag{2.21}$$

by convexity of $2^x$. We say in this case that $(a, b)$ is more *balanced* than $(a - \varepsilon, b + \varepsilon)$.

## 2.7 Exponential Time Subroutines

So far, we analyzed the running time of Algorithm **mis** for graphs of maximum degree 5 only. In this section, we see one way to combine different analyses for different algorithms or different analyses of the same algorithm. For an algorithm that can be divided into different stages, the following lemma shows that it is allowed to "decrease" the measure when passing from one stage to another.

**Lemma 2.6.** *Let $A$ be an algorithm for a problem $P$, $B$ be an algorithm for (special instances of) $P$, $c \geq 0$ be a constant, and $\mu(\cdot), \mu'(\cdot), \eta(\cdot)$ be measures for the instances of $P$, such that for any input instance $I$, $\mu'(I) \leq \mu(I)$ and for any input instance $I$, $A$ either solves $P$ on $I$ by invoking $B$ with running time $\mathcal{O}(\eta(I)^{c+1})2^{\mu'(I)}$, or reduces $I$ to instances $I_1, \ldots, I_k$, solves these recursively, and combines their solutions to solve $I$, using time $\mathcal{O}(\eta(I)^c)$ for the reduction and combination steps (but not the recursive solves),*

$$(\forall i) \quad \eta(I_i) \leq \eta(I) - 1, \text{ and} \tag{2.22}$$

$$\sum_{i=1}^{k} 2^{\mu(I_i)} \leq 2^{\mu(I)}. \tag{2.23}$$

*Then $A$ solves any instance $I$ in time $\mathcal{O}(\eta(I)^{c+1})2^{\mu(I)}$.*

*Proof.* Again, the result follows by induction on $\eta(I)$. For the base case, we assume that the algorithm returns the solution to an empty instance in time $\mathcal{O}(1)$. If an instance $I$ is solved in time $\mathcal{O}(\eta(I)^{c+1})2^{\mu'(I)}$ by invoking Algorithm $B$, then the running time of algorithm $A$ to solve instance $I$ is $T_A(I) = \mathcal{O}(\eta(I)^{c+1})2^{\mu'(I)} = \mathcal{O}(\eta(I)^{c+1})2^{\mu(I)}$ as $\mu'(I) \leq \mu(I)$. Otherwise,

$$T_A(I) = \mathcal{O}(\eta(I)^c) + \sum_{i=1}^{k} T_A(I_i) \qquad \text{(by definition)}$$

$$= \mathcal{O}(\eta(I)^{c+1})2^{\mu(I)}. \qquad \text{(following the proof of Lemma 2.5)}$$

$\square$

Very related is the idea of a piecewise linear measure [GSB95, DJW05, FK05, Wah08] which can be seen as dividing the algorithm in different subroutines — one for each linear piece of the measure.

To derive an upper bound for the worst case running time of Algorithm **mis** on general instances, use Lemma 2.6 with $A = B = $ **mis**, $c = 2$, $\mu(G) = 0.35805n$, $\mu'(G) = \sum_{i=1}^{5} w_i n_i$ with the values of the $w_i$'s as in Table 2.3 on page 40, and $\eta(G) = n$. For every instance $G$, $\mu'(G) \leq \mu(G)$ because for each $i \in \{1, \dots, 5\}$, $w_i \leq 0.35805$. Further, for each $d \geq 6$,

$$(0.35805, (d+1) \cdot 0.35805) \leq 1.$$

Thus, Algorithm **mis** has running time $\mathcal{O}(1.2817^n)$ for graphs of arbitrary degrees.

## 2.8 Towards a Tighter Analysis

The previous sections of this chapter provide us with fairly good tools to upper bound the worst case running times of branching algorithms. This section shows how one can enhance the measures to squeeze the upper bound a little bit closer to the actual worst case running time of the algorithm.

### 2.8.1 Structures that Arise Rarely

It was — to our knowledge — first observed by Robson [Rob86] that branching on a local configuration only affects the overall running time of the algorithm by a constant factor if this local configuration is only selected (for branching) a constant number of times on the path from a leaf to the root of any search tree corresponding to an execution of the algorithm.

This can be formally proved by slightly modifying the measure of the instance. Let $C$ be a constant and $s$ be an "undesired" local configuration which may only be selected once on the path from a leaf to the root in any search tree of the algorithm. Let

$$\mu'(I) := \begin{cases} \mu(I) + C & \text{if } s \text{ may be selected in the current subtree, and} \\ \mu(I) & \text{otherwise.} \end{cases} \qquad (2.24)$$

Consider an instance $I$ where $s$ is selected for branching. Then $\mu'(I) = \mu(I) + C$ and for each subinstance $I_i, 1 \leq i \leq k$, $\mu'(I_i) = \mu(I_i)$. By giving a high enough constant value to $C$, the branching number

$$\big(\mu(I) - \mu(I_1) + C, \dots, \mu(I) - \mu(I_k) + C\big)$$

is at most 1, with the reasonable assumption that branching on $s$ does not increase the measure

> **else**
>     Select $v \in V$ such that
>         (1) $v$ has maximum degree, and
>         (2) among all vertices satisfying (1), $v$ has a neighbor of minimum degree
>     **return** max $(1 + \mathbf{mis}(G \setminus N[v]), \mathbf{mis}(G \setminus v))$

Figure 2.7: Modified branching rule for Algorithm **mis**

| $i$ | $w_i$ | $h_i$ |
|---|---|---|
| 0 | 0 | 0 |
| 1 | 0 | 0 |
| 2 | 0.207137 | 0.207137 |
| 3 | 0.322203 | 0.115066 |
| 4 | 0.343587 | 0.021384 |
| 5 | 0.347974 | 0.004387 |

Table 2.4: An optimal assignment of the weights for the measure $\mu(G) = \sum_{i=1}^{5} w_i n_i$ for the analysis of Algorithm **mis** modified according to Figure 2.7

of the subinstances by more than a constant. The overall running time in Lemma 2.5 on page 37 can then be upper bounded by $\eta(I)^{c+1} 2^{\mu'(I)} = \eta(I)^{c+1} 2^{\mu(I)+C} = \eta(I)^{c+1} 2^C \cdot 2^{\mu(I)}$.

This argument can easily be iterated for undesired local configurations that only arise at most a constant number of times on the path from a leaf to the root of any search tree corresponding to an execution of the algorithm.

Let us slightly modify the selection of the local configuration for the second branching rule of Algorithm **mis** on page 30 as shown in Figure 2.7.

With this modification, Algorithm **mis** only selects $v$ of degree $d$ with all neighbors of degree $d$ when the graph is $d$-regular. As no connected $d$-regular graph contains any other $d$-regular subgraph, let us define

$$\mu'(G) = \mu(G) + \sum_{d=3}^{5} \mathsf{K}_\delta(G \text{ has a } d\text{-regular subgraph}) C_d \qquad (2.25)$$

where $C_d, 3 \leq d \leq 5$, are constants.

A little care is now required for analyzing the first branching rule of Algorithm **mis**, as we do not necessarily have that $\mu'(G) = \mu'(G_1) + \mu'(G \setminus V(G_1))$. Suppose $\mu(G_1), \mu(G \setminus V(G_1)) > K$ for a large enough constant $K$. Then the constraint

$$2^{\mu'(G_1)} + 2^{\mu'(G \setminus V(G_1))} \leq 2^{\mu'(G)}$$

is satisfied. Otherwise, $\mu(G_1) \leq K$ or $\mu(G \setminus V(G_1)) \leq K$ and a maximum independent set can be computed in constant time for the subgraph whose measure $\mu$ is bounded by $K$.

Turning to the second branching rule, all the branching numbers for regular instances can now be ignored as they are irrelevant to the worst case behavior of the algorithm. Thus, we obtain the following set of branching numbers. For each $d, 3 \leq d \leq 5$ and all $p_i, 2 \leq i \leq d$ such that $\sum_{i=2}^{d} p_i = d$ and $p_d \neq d$,

$$\left( w_d + \sum_{i=2}^{d} p_i \cdot (w_i - w_{i-1}), w_d + \sum_{i=2}^{d} p_i \cdot w_i + h_d \right).$$

All these branching numbers are at most 1 with the optimal set of weights in Table 2.4. Thus, Algorithm **mis**, modified according to Figure 2.7 on the preceding page, has running time $\mathcal{O}(1.2728^n)$.

With the same arguments, we may also loosen one condition of Lemma 2.6 on page 42; namely, we can replace $\mu'(I) \leq \mu(I)$ by $\mu'(I) \leq \mu(I) + C$ for a constant $C$.

## 2.8.2   State Based Measures

Sometimes it is very useful to introduce states of the algorithm in order to specify that some branching with a "bad" branching number is always followed by a branching with a "better" branching number. A convenient way to amortize over the branching numbers is to add a constant to the measure depending on some properties of the instance, see for example [Wah04, CKX05] or Chapters 4 and 5 for applications.

More formally, the measure of an instance $I$ is divided in two parts

$$\mu'(I) := \mu(I) + \Psi(I),$$

where $\Psi : \mathcal{I} \rightarrow \mathbb{R}^+$ is a function from the set of instances $\mathcal{I}$ to the positive reals depending on global properties of the instance. Being additive and constant–bounded, the function $\Psi(\cdot)$ increases the running time only by a constant factor and has the potential to decrease the branching numbers.

The intuition is that $\Psi(I)$ is larger for branchings whose branching numbers with respect to $\mu$ are high, but that create subinstances for which the branchings have low branching numbers with respect to $\mu$. In this way, the function $\Psi(\cdot)$ may enable us to decrease the highest branching numbers by increasing some branching numbers that are not tight.

Consider Algorithm **mis2** in Figure 2.8. It is similar to Algorithm **mis**, with the enhancements discussed in Section 2.2 and Subsection 2.8.1, but it also contains the folding rule for vertices of degree 2 (lines 8–9). For a graph $G = (V, E)$ and a vertex $v \in V$ of degree 2 whose neighbors are not adjacent, the operation $\mathbf{fold}(G, v)$ returns the graph obtained from $G \setminus N[v]$ by adding a new vertex $f_v$ and making it adjacent to the vertices in $N_G(v)$. Note that the algorithm does not fold degree-2 vertices whose neighbors are adjacent due to the dominance rule in lines 3–4. The folding operation has been used in [Bei99, CKJ01, FGK09b] and the following lemma is well-known.

**Lemma 2.7.** *Let $G = (V, E)$ be a graph and $v \in V$ be a vertex of degree 2 whose neighbors are not adjacent. Then, $\alpha(G) = \alpha(\mathbf{fold}(G, v)) + 1$, where $\alpha(\cdot)$ denotes the size of a maximum independent set.*

---

Algorithm **mis2**$(G)$
**Input**   : A graph $G = (V, E)$.
**Output**: The size of a maximum independent set of $G$.

1 **if** $\Delta(G) \leq 2$ **then**                                    // `G has maximum degree at most 2`
2 | **return** the size of a maximum independent set of $G$ in polynomial time

3 **else if** $\exists u, v \in V : N[u] \subseteq N[v]$ **then**                        // `u dominates v`
4 | **return** **mis2**$(G \setminus v)$

5 **else if** $G$ is not connected **then**
6 | Let $G_1$ be a connected component of $G$
7 | **return** **mis2**$(G_1)$ + **mis2**$(G \setminus V(G_1))$

8 **else if** $\exists v \in V : d(v) = 2$ **then**                         // `fold vertices of degree 2`
9 | **return** $1$ + **mis2**$(\textbf{fold}(G, v))$

10 **else**
11 | Select $v \in V$ such that
   |    (1) $v$ has maximum degree, and
   |    (2) among all vertices satisfying (1), $v$ has a neighbor of minimum degree
12 | **return** $\max\left(1 + \textbf{mis2}(G \setminus N[v]), \textbf{mis2}(G \setminus v)\right)$

Figure 2.8: Algorithm **mis2**$(G)$, computing the size of a maximum independent set of any input graph $G$

---

*Proof.* To show that $\alpha(G) \leq \alpha(\textbf{fold}(G, v)) + 1$, let $I$ be a maximum independent set of $G$. If $N_G[v] \cap I$ contains only one vertex $x$, then $I \setminus \{x\}$ is an independent set of $\textbf{fold}(G, v)$ of size $\alpha(G) - 1$. Otherwise, $N_G[v] \cap I = N_G(v)$, and thus $(I \setminus N_G(v)) \cup \{f_v\}$ is an independent set of $\textbf{fold}(G, v)$ of size $\alpha(G) - 1$.

To show that $\alpha(G) \geq \alpha(\textbf{fold}(G, v)) + 1$, let $I$ be a maximum independent set of $\textbf{fold}(G, v)$. If $f_v \notin I$, then $I \cup \{v\}$ is an independent set of $G$ of size $\alpha(\textbf{fold}(G, v)) + 1$. Otherwise, $f_v \in I$, and then $(I \setminus \{f_v\}) \cup N_G(v)$ is an independent set of $G$ of size $\alpha(\textbf{fold}(G, v)) + 1$.                                □

The folding operation of Algorithm **mis2** is a simplification rule removing vertices of degree 2. On the positive side, this means that in the analysis of the branching in lines 10–12, we may assume that the selected vertex has no neighbors of degree 2. On the negative side, the folding operation has the potential to create regular instances over and over again. So, we cannot just ignore branchings on regular instances as in the previous subsection.

To analyze Algorithm **mis2**, let us use $\Psi(G) := \mathsf{K}_\delta(G$ is $d$-regular for some $d \geq 4) \cdot R$ for a constant $R$, and add it to our measure $\mu$. Note that the measure of an instance increases by $R$ when a branching on a nonregular graph creates a regular graph, and decreases by $R$ when the instance was regular and becomes nonregular. This decreases the branching numbers for regular instances and increases the branching numbers for nonregular instances, and thereby amortizes over the branching numbers. When branching on a non-regular instance, the created subinstances may be regular, in which case the measure decrease is smaller by $R$. Thus, for each $d, 4 \leq d \leq 6$ and all $p_i, 3 \leq i \leq d$ such that $\sum_{i=3}^{d} p_i = d$ and $p_d \neq d$, we have the branching

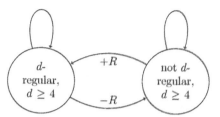

Figure 2.9: A state graph for the analysis of Algorithm **mis2**

number

$$\left(w_d - R + \sum_{i=3}^{d} p_i \cdot (w_i - w_{i-1})\right), w_d - R + \sum_{i=3}^{d} p_i \cdot w_i + h_d\right).$$

A branching on a vertex of degree 3 (the instance is 3-regular) does not create a $d$-regular instance with $d \geq 4$ unless the graph has constant size:

$$(w_3 + 3 \cdot (w_3 - w_2), 4w_4 + h_3).$$

A branching on a vertex $v$ of a $d$-regular instance with $d \geq 4$ creates a non-regular instance when $v$ is deleted and may create a $d$-regular instance when $N[v]$ is deleted. For each $d \geq 4$, we obtain the branching number

$$(w_d + R + d \cdot (w_d - w_{d-1}), (d+1) \cdot w_d + h_d).$$

Computing the optimal weights (which gives $R \simeq 0.016$) such that all branching numbers are at most 1, we find that Algorithm **mis2** has running time $\mathcal{O}(1.2571^n)$.

## 2.9 Conclusion

In this chapter, we have seen how to establish worst case upper bounds on the running time of branching algorithms. Methods and ideas how to improve the analysis have been extensively discussed and exemplified on an algorithm for MAXIMUM INDEPENDENT SET, which served as an introductory example; the goal was to illustrate methods of analysis, and not to design a faster algorithm for this problem.

The methods and ideas presented in this chapter will be used in the forthcoming chapters to design and analyze competitive algorithms for various problems.

On a side note, measure bases analyses have also been successfully used to design parameterized algorithms [FGR09, Gas09, LS09].

# Chapter 3

# Feedback Vertex Sets

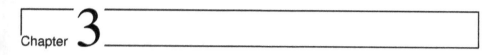

In all things there is a law of cycles.

*Publius Cornelius Tacitus*

In this chapter we present an $\mathcal{O}(1.7548^n)$ time algorithm finding a minimum feedback vertex set in an undirected graph on $n$ vertices. We also prove that a graph on $n$ vertices can contain at most $1.8638^n$ minimal feedback vertex sets and that there exist graphs having $105^{n/10} \approx 1.5926^n$ minimal feedback vertex sets. The optimization algorithm, as well as the upper bound on the number of minimal feedback vertex sets use a measure based analysis as presented in Section 2.3. The lower bound on the number of minimal feedback vertex sets is derived via the construction of an infinite family of graphs with this number of minimal feedback vertex sets.

## 3.1 Motivation and Related Work

The problem of finding a minimum feedback vertex set in a graph, that is the smallest set of vertices whose removal makes the graph acyclic, has many applications, for example in genome sequence assembly [PKS04] and VLSI chip design [KVZ01]. Its history can be traced back to the early '60s (see the survey of Festa et al. [FPR99]). It is also one of the classical $\mathcal{NP}$-complete problems from Karp's list [Kar72]. Thus not surprisingly, for several decades, many different algorithmic approaches were tried on this problem including approximation algorithms [BBF99, BYGNR98, ENSZ00, KK01], linear programming [CGHW98], local search [BMT00], polyhedral combinatorics [CDZ02, FR96], probabilistic algorithms [PQR99], parameterized algorithms [DF99, GGH$^+$06, DFL$^+$07, CFL$^+$08], and kernelization [BECF$^+$06, Bod07, Tho09].

The problem is approximable within a factor of 2 in polynomial time [BBF99]. It was also extensively studied from the point of view of parameterized complexity. There was a chain of improvements (see for example [RSS06]) concluding with two $2^{\mathcal{O}(k)}n^{\mathcal{O}(1)}$-time algorithms obtained independently by different research groups [DFL$^+$05, GGH$^+$06]. It had been open for a long time whether computing a feedback vertex set of a *directed* graph is fixed-parameter tractable. Recently, this question has been resolved positively [CLL$^+$08].

Although the topic of exact exponential time algorithms for $\mathcal{NP}$-hard problems has led to much research in recent years, and despite much progress on exponential time solutions to other graph problems such as CHROMATIC NUMBER [Bys04a, BHK09], MAXIMUM INDEPENDENT SET [FGK09b, Rob86], and MINIMUM DOMINATING SET [FGK09b], no algorithm faster than the trivial $\mathcal{O}^*(2^n)$ was known for FEEDBACK VERTEX SET until 2006. For some special graph classes, like bipartite graphs or graphs of maximum degree 4, algorithms of running time $\mathcal{O}(1.8621^n)$ and $\mathcal{O}(1.945^n)$ respectively can be found in the literature [FP05, RSS05]. Further, an algorithm for general graphs with running time $\mathcal{O}(1.7347^n)$ has been obtained by Fomin and Villanger [FV10].

## 3.2   Discussion of Results

The exact algorithm presented here solves FEEDBACK VERTEX SET in time $\mathcal{O}(1.7548^n)$. The main idea behind breaking the $2^n$ barrier for FEEDBACK VERTEX SET is based on the choice of the measure of the subproblems recursively generated by the algorithm.

By making use of similar ideas, we show that every graph on $n$ vertices contains at most $1.8638^n$ minimal feedback vertex sets. It is the first known upper bound for the number of minimal feedback vertex sets breaking the trivial $\mathcal{O}(2^n/\sqrt{n})$ bound (which is roughly the maximum number of subsets of an $n$-element set such that none of them is contained in another one). This bound has algorithmic consequences as well. By the result of Schwikowski and Speckenmeyer [SS02], all minimal feedback vertex sets can be enumerated with polynomial time delay. Thus our result implies that the running time of the algorithm by Schwikowski and Speckenmeyer is $\mathcal{O}(1.8638^n)$. We also show that there exist graphs with at least $1.5926^n$ minimal feedback vertex sets.

The rest of this chapter is organized as follows. Section 3.3 contains preliminary results. In Section 3.4 we present an $\mathcal{O}(1.7548^n)$ time algorithm finding a minimum feedback vertex set in a graph on $n$ vertices. In Section 3.5 we prove that every graph on $n$ vertices has at most $1.8638^n$ minimal feedback vertex sets and that there exists an infinite family of graphs having $1.5926^n$ minimal feedback vertex sets.

## 3.3   Preliminaries

The set $V'$ is a feedback vertex set if and only if $G \setminus V'$ is a forest. A feedback vertex set is *minimal* if it does not contain any other feedback vertex set as a proper subset, and *minimum* if it has minimum cardinality among all feedback vertex sets in a graph. Let us note that $X$ is a minimal (minimum) feedback vertex set if and only if $G \setminus X$ is a maximal (maximum) induced forest. Thus the problem of finding a minimum feedback vertex set is equivalent to the problem of finding a maximum induced forest. Similarly, the number of minimal feedback vertex sets in a graph is equal to the number of maximal induced forests. For the description of the algorithm it is more convenient to work with maximum induced forests than with feedback vertex sets.

We call a subset $F \subseteq V$ *acyclic* if $G[F]$ is a forest. If $F$ is acyclic then every connected component of $G[F]$ on at least two vertices is called *non-trivial*. If $T$ is a non-trivial connected component of $G[F]$ then we denote by $\mathrm{Id}(T, t)$ the operation of contracting all edges of $T$ into one vertex $t$ and removing appeared loops. Note that this operation may create multiedges in $G$. We denote by $\mathrm{Id}^*(T, t)$ the operation $\mathrm{Id}(T, t)$ followed by the removal of all vertices connected with $t$ by multiedges.

For an acyclic subset $F \subseteq V$, denote by $\mathcal{M}_G(F)$ and by $\mathcal{M}_G^*(F)$ the set of all maximal and maximum acyclic supersets of $F$ in $G$, respectively (we omit the subindex $G$ when it is clear from the context which graph is meant). Let $\mathcal{M}^* := \mathcal{M}^*(\emptyset)$. Then the problem of finding a maximum induced forest can be stated as finding an element of $\mathcal{M}^*$. We solve a more general problem, namely finding an element of $\mathcal{M}^*(F)$ for an arbitrary acyclic subset $F$.

To simplify the description of the algorithm, we suppose that $F$ is always an independent set. The next lemma justifies this assumption.

**Lemma 3.1.** *Let $G = (V, E)$ be a graph, $F \subseteq V$ be an acyclic subset of vertices and $T$ be a non-trivial connected component of $G[F]$. Denote by $G'$ the graph obtained from $G$ by the operation $\mathrm{Id}^*(T, t)$ and let $F' := F \cup \{t\} \setminus T$. Then*

- $X \in \mathcal{M}_G(F)$ *if and only if* $X' \in \mathcal{M}_{G'}(F')$, *and*

- $X \in \mathcal{M}_G^*(F)$ *if and only if* $X' \in \mathcal{M}_{G'}^*(F')$,

*where $X' := X \cup \{t\} \setminus T$.*

*Proof.* Assume that $X \in \mathcal{M}_G(F)$. If after the operation $\mathrm{Id}(T, t)$ a vertex $v$ is connected with $t$ by a multiedge, then the set $T \cup \{v\}$ is not acyclic in $G$. Hence, no element of $\mathcal{M}_G(F)$ may contain $v$. In other words, $X$ does not contain any vertices removed by the transformation from $G$ to $G'$ and hence $X' = X \cup \{t\} \setminus T$ is a set of vertices of $G'$. Moreover, $X'$ is an acyclic subset of $G'$. To see this, assume by contradiction that $X'$ induces a cycle $C'$ in $G'$. Then $C'$ necessarily includes $t$ because otherwise $C'$ is induced by $X$ in $G$ in contradiction to the acyclicity of $X$. Let $x_1$ and $x_2$ be the two neighbors of $t$ in $C'$. It follows that there is a path in $G$ from $x_1$ to $x_2$ including vertices of $T$ only. Replace $t$ in $C'$ by such a path. As a result we obtain a cycle induced by $X$ in $G$ in contradiction to the acyclicity of $X$. It remains to show that $X'$ is a maximal acyclic subset of $G'$. For this purpose, assume that there is a vertex $v \in V(G') \setminus X'$ such that $X' \cup \{v\}$ is an acyclic subset. Then $X \cup \{v\}$ is an acyclic subset of $G$ (any cycle in $X \cup \{v\}$ can be transformed into a cycle in $X' \cup \{v\}$ by the operation $\mathrm{Id}(T, t)$) larger than $X$ in contradiction to the maximality of $X$.

Arguing similarly, we can prove that if $X' \in \mathcal{M}_{G'}(F')$ then $X \in \mathcal{M}_G(F)$ and that $X \in \mathcal{M}_G^*(F)$ if and only if $X' \in \mathcal{M}_{G'}^*(F')$. $\square$

By using the operation $\mathrm{Id}^*$ on every non-trivial component of $F$, we obtain an independent set $F'$.

The following lemma is used to justify the main branching rule of the algorithm.

**Lemma 3.2.** *Let $G = (V, E)$ be a graph, $F \subseteq V$ be an independent subset of vertices and $v \notin F$ be a vertex adjacent to exactly one vertex $t \in F$. Then*

1. For every $X \in \mathcal{M}(F)$, either $v$ or at least one vertex of $N(v) \setminus \{t\}$ is in $X$.

2. There exists $X \in \mathcal{M}^*(F)$ such that either $v$ or at least two vertices of $N(v) \setminus \{t\}$ are in $X$.

*Proof.* 1. If there is $X \in \mathcal{M}(F)$ such that $v \notin X$ and no vertex of $N(v) \setminus \{t\}$ is in $X$, then $X \cup \{v\}$ is also an induced forest of $G$. Thus $X$ is not maximal, which is a contradiction.

2. Let us consider $X \in \mathcal{M}^*(F)$ such that $v \notin X$. By item 1, at least one vertex $z \in N(v) \setminus \{t\}$ is in $X$. For the sake of contradiction, let us assume that $z$ is the only such vertex. Since $X$ is maximal, we have that $X \cup \{v\}$ is not acyclic. Because $v$ is of degree at most 2 in $G[X \cup \{v\}]$, we conclude that all the cycles in $G[X \cup \{v\}]$ must contain $z$. Then the set $X \cup \{v\} \setminus \{z\}$ is in $\mathcal{M}^*(F)$ and satisfies the conditions.                                                    $\square$

Consequently, if $N(v) = \{t, v_1, v_2, \ldots, v_k\}$, then there exists $X \in \mathcal{M}^*(F)$ satisfying one of the following properties:

1. $v \in X$;

2. $v \notin X$, $v_i \in X$ for some $i \in \{1, 2, \ldots, k-2\}$ while $v_j \notin X$ for all $j < i$;

3. $v, v_1, v_2, \ldots, v_{k-2} \notin X$ but $v_{k-1}, v_k \in X$.

In particular, if $k \leq 1$, then $v \in X$ for some $X \in \mathcal{M}^*(F)$.

The following lemma is needed to handle the case where every vertex in $V \setminus F$ is adjacent to a vertex $t \in F$. We reduce this case to finding a maximal (respectively maximum) independent set in the graph $G[V \setminus F]$ with some additional edges.

**Lemma 3.3.** *Let $G = (V, E)$ be a graph and $F$ be an independent set in $G$ such that $V \setminus F = N(t)$ for some $t \in F$. Consider the graph $G' := G[N(t)]$ and for every pair of vertices $u, v \in N(t)$ having a common neighbor in $F \setminus \{t\}$ add an edge $uv$ to $G'$. Denote the obtained graph by $H$ and let $I \subseteq N(t)$. Then $F \cup I \in \mathcal{M}_G(F)$ if and only if $I$ is a maximal independent set in $H$. In particular, $F \cup I \in \mathcal{M}_G^*(F)$ if and only if $I$ is a maximum independent set in $H$.*

*Proof.* Let $X \in \mathcal{M}_G(F)$ and $u, v \in V \setminus F$. If $uv \in E$ then $u, v, t$ form a triangle. If there is a vertex $w \in F \setminus \{t\}$ adjacent to both $u$ and $v$ then $tuwv$ is a 4-cycle. In both cases, $X$ cannot contain $u$ and $v$ at the same time. On the other hand, if $I \subseteq N(t)$ such that no two vertices of $I$ are adjacent in $G$ and no two vertices of $I$ have a common neighbor except $t$ then $F \cup I$ induces a forest in $G$. Therefore, $X \in \mathcal{M}_G(F)$ if and only if $X \setminus F$ is a maximal independent set in $H$.                                                    $\square$

There are several fast exponential algorithms computing a maximum independent set in a graph. We use the polynomial space algorithm of Kneis et al.

**Theorem 3.4** ([KLR09]). *Let $G$ be a graph on $n$ vertices. Then a maximum independent set in $G$ can be found in time $\mathcal{O}(2^{0.2789n})$ and polynomial space.*

For the upper bound on $|\mathcal{M}_G(\emptyset)|$ for any graph $G$, we also need the following well known result of Moon and Moser [MM65].

**Theorem 3.5** ([MM65]). *A graph on $n$ vertices has at most $3^{n/3}$ maximal independent sets.*

## 3.4 Computing a Minimum Feedback Vertex Set

In this section we show how to compute the minimum size of a feedback vertex set. Our algorithm can easily be turned into an algorithm computing at least one such set. Instead of working with feedback vertex sets directly, the algorithm finds the maximum size of an induced forest in a graph. In fact, it solves a more general problem: for any acyclic set $F$ it finds the maximum size of an induced forest containing $F$.

During the work of the algorithm one vertex $t \in F$ is called an *active vertex*. The algorithm branches on a chosen neighbor of $t$. Let $v \in N(t)$. Denote by $K$ the set of all vertices of $F$ other than $t$ that are adjacent to $v$. Let $G'$ be the graph obtained after the operation $\mathrm{Id}(K \cup \{v\}, u)$. We say that a vertex $w \in V \setminus \{t\}$ is a *generalized neighbor* of $v$ in $G$ if $w$ is a neighbor of $u$ in $G'$. Denote by $\mathrm{gd}(v)$ the *generalized degree* of $v$ which is the number of its generalized neighbors.

The description of the algorithm consists of a sequence of cases and subcases. To avoid a confusing nesting of if-then-else statements let us use the following convention: the first case which applies is used in the algorithm. Thus, inside a given case, the hypotheses of all previous cases are assumed to be false.

Algorithm $\mathrm{mif}(G, F)$ computing for a given graph $G$ and an acyclic set $F$ the maximum size of an induced forest containing $F$ is described by the following preprocessing and main procedures (let us note that $\mathrm{mif}(G, \emptyset)$ computes the maximum size of an induced forest in $G$).

**Preprocessing**

1. If $G$ consists of $k \geq 2$ connected components $G_1, G_2, \ldots, G_k$, then the algorithm is called on each of the components and

$$\mathrm{mif}(G, F) = \sum_{i=1}^{k} \mathrm{mif}(G_i, F_i),$$

where $F_i := V(G_i) \cap F$ for all $i \in \{1, 2, \ldots, k\}$.

2. If $F$ is not independent, then apply operation $\mathrm{Id}^*(T, v_T)$ on an arbitrary non-trivial component $T$ of $F$. If $T$ contains the active vertex then $v_T$ becomes active. Let $G'$ be the resulting graph and let $F'$ be the set of vertices of $G'$ obtained from $F$. Then

$$\mathrm{mif}(G, F) = \mathrm{mif}(G', F') + |T| - 1.$$

**Main procedures**

1. If $F = V$ then $\mathcal{M}_G(F) = \{V\}$. Thus,

$$\mathrm{mif}(G, F) = |V|.$$

2. If $F = \emptyset$ and $\Delta(G) \leq 1$ then $\mathcal{M}_G(F) = \{V\}$ and

$$\mathrm{mif}(G, F) = |V|.$$

3. If $F = \emptyset$ and $\Delta(G) \geq 2$ then the algorithm chooses a vertex $t$ in $G$ of degree at least 2. Then $t$ is either contained in a maximum induced forest or not. Thus the algorithm branches on two subproblems and returns the maximum:

$$\mathtt{mif}(G, F) = \max\{\mathtt{mif}(G, F \cup \{t\}),$$
$$\mathtt{mif}(G \setminus \{t\}, F)\}.$$

4. If $F$ contains no active vertex then choose an arbitrary vertex $t \in F$ as an active vertex. Denote the active vertex by $t$ from now on.

5. If $V \setminus F = N(t)$ then the algorithm constructs the graph $H$ from Proposition 3.3 and computes a maximum independent set $I$ in $H$. Then

$$\mathtt{mif}(G, F) = |F| + |I|.$$

6. If there is $v \in N(t)$ with $\mathrm{gd}(v) \leq 1$ then add $v$ to $F$:

$$\mathtt{mif}(G, F) = \mathtt{mif}(G, F \cup \{v\}).$$

7. If there is $v \in N(t)$ with $\mathrm{gd}(v) \geq 4$ then either add $v$ to $F$ or remove $v$ from $G$:

$$\mathtt{mif}(G, F) = \max\{\mathtt{mif}(G, F \cup \{v\}),$$
$$\mathtt{mif}(G \setminus \{v\}, F)\}.$$

8. If there is $v \in N(t)$ with $\mathrm{gd}(v) = 2$ then denote its generalized neighbors by $w_1$ and $w_2$. Either add $v$ to $F$ or remove $v$ from $G$ but add $w_1$ and $w_2$ to $F$. If adding $w_1$ and $w_2$ to $F$ induces a cycle, we just ignore the last branch.

$$\mathtt{mif}(G, F) = \max\{\mathtt{mif}(G, F \cup \{v\}),$$
$$\mathtt{mif}(G \setminus \{v\}, F \cup \{w_1, w_2\})\}.$$

9. If all vertices in $N(t)$ have exactly three generalized neighbors then at least one of these vertices must have a generalized neighbor outside $N(t)$, since the graph is connected and the condition of the case Main 5 does not hold. Denote such a vertex by $v$ and its generalized neighbors by $w_1$, $w_2$ and $w_3$ in such a way that $w_1 \notin N(t)$. Then we either add $v$ to $F$; or remove $v$ from $G$ but add $w_1$ to $F$; or remove $v$ and $w_1$ from $G$ and add $w_2$ and $w_3$ to $F$. Similarly to the previous case, if adding $w_2$ and $w_3$ to $F$ induces a cycle, we just ignore the last branch.

$$\mathtt{mif}(G, F) = \max\{\mathtt{mif}(G, F \cup \{v\}),$$
$$\mathtt{mif}(G \setminus \{v\}, F \cup \{w_1\}),$$
$$\mathtt{mif}(G \setminus \{v, w_1\}, F \cup \{w_2, w_3\})\}.$$

The correctness and the running time of the algorithm are analyzed in the following.

**Theorem 3.6.** *Let $G$ be a graph on $n$ vertices. Then a maximum induced forest of $G$ can be found in time $\mathcal{O}(1.7548^n)$.*

*Proof.* Let us consider Algorithm $\mathtt{mif}(G, F)$ described above. The correctness of **Preprocessing 1** and **Main 1, 2, 3, 4, 7** is clear. The correctness of **Main 5** follows from Lemma 3.3, while the correctness of **Preprocessing 2** and **Main 6, 8, 9** follows from Lemma 3.1 and 3.2 (indeed, applying Lemma 3.2 to the vertex $u$ of the graph $G'$ shows that for some $X \in \mathcal{M}_G(F)$ either $v$ or at least two of its generalized neighbors are in $X$).

In order to evaluate the time complexity of the algorithm we use Lemma 2.6 with the following measures:

$$\mu(G, F, t) := \alpha|N(t)| + \beta|V \setminus (F \cup N(t))|$$
$$\mu'(G, F) := 0.2789|V \setminus F|$$
$$\eta(G, F) := |V| + |V \setminus F|$$

with $\alpha := 0.415$ and $\beta := 0.8113$. In other words, for the measure $\mu$ each vertex in $F$ has weight 0, each vertex in $N(t)$ has weight $\alpha$, each other vertex has weight $\beta$, and $\mu$ is equal to the sum of the vertex weights. We will prove that a problem of size $\mu$ can be solved in time $\mathcal{O}(2^\mu)$. As $\mu \leq \beta(n - |F|)$, the running time is $\mathcal{O}(1.7548^{n-|F|})$.

First, we prove that every simplification and every branching which reduces an instance $(G, F)$ to an instance $(G_1, F_1)$, the measure $\eta(G, F)$ decreases by at least 1, that is $\eta(G_1, F_1) \leq \eta(G, F) - 1$.

For the preprocessing cases and the cases **Main 3, 6, 7, 8, 9**, this immediately follows from the description. Cases **Main 1, 2** do not make any recursive calls. As case **Main 4** never occurs in two consecutive nodes of the search tree, its statement may be reformulated as "choose an arbitrary vertex $t$ as new active vertex and go through the list of cases again to select the appropriate one". That is, the node corresponding to case **Main 4** may be analyzed together with the next node where $t$ is specified. Finally in case **Main 5**, $\mu' \leq \mu$ for every instance and by Theorem 3.4, a maximum independent set in $H$ can be found in time $\mathcal{O}(2^{\mu'})$.

It is now clear that the following steps do not increase the measure $\mu$ and do not contribute to the exponential factor of the running time of the algorithm: **Preprocessing 1, 2** and **Main 1, 2, 4, 6**.

In all the remaining cases the algorithm is called recursively on smaller instances. We consider these cases separately.

In case **Main 3** every vertex has weight $\beta$. So, removing $v$ leads to an instance of size $\mu - \beta$. Otherwise, $v$ becomes active after the next Main 4 step. Then all its neighbors will have weight $\alpha$, and we obtain an instance of size at most $\mu - \beta - 2(\beta - \alpha)$ since $v$ has degree at least 2. Thus the branching number of this case is at most

$$(\beta, 3\beta - 2\alpha) \leq 1.$$

In case **Main 7** removing vertex $v$ decreases the size of the instance by $\alpha$. If $v$ is added to $F$ then we obtain a non-trivial component in $F$, which is contracted into a new active vertex $t'$ at the next Preprocessing 2 step. Those of the generalized neighbors of $v$ that had weight $\alpha$ will be connected with $t'$ by multiedges and thus removed during the next Preprocessing 2 step. If a generalized neighbor of $v$ had weight $\beta$ then it will become a neighbor of $t'$, that is its weight becomes $\alpha$. Thus, in any case the size of the instance decreases by at least $\alpha + 4(\beta - \alpha)$ as $\beta - \alpha < \alpha$. So, we have a branching number of at most

$$(\alpha, 4\beta - 3\alpha) \leq 1.$$

In case **Main 8** we distinguish three subcases depending on the weights of the generalized neighbors of $v$. Let $i$ be the number of generalized neighbors of $v$ having weight $\beta$. Adding $v$ to $F$ reduces the weight of a generalized neighbor either from $\alpha$ to $0$ or from $\beta$ to $\alpha$. Removing $v$ from the graph reduces the weight of both generalized neighbors of $v$ to $0$ (since we add them to $F$). According to this, we obtain the following branching numbers: for $i \in \{0, 1, 2\}$,

$$(\alpha + i \cdot (\beta - \alpha) + (2 - i) \cdot \alpha, \alpha + i \cdot \beta + (2 - i) \cdot \alpha) \leq 1.$$

Case **Main 9** is considered analogously to Main 8, except that at least one of the generalized neighbors of $v$ has weight $\beta$, that is $i \geq 1$ ($i = 0$ is excluded by Main 5). In this case, we have for $i \in \{1, 2, 3\}$,

$$(\alpha + i \cdot (\beta - \alpha) + (3 - i) \cdot \alpha, \alpha + \beta, \alpha + i \cdot \beta + (3 - i) \cdot \alpha) \leq 1.$$

Thus all the branching numbers are at most 1 and the proof follows from Lemma 2.6. □

*Remark* 3. The only tight constraint is the one of Case Main 7 when $v$ has generalized degree 4. Thus, an improvement of this case would improve the overall (upper bound of the) running time of the algorithm.

## 3.5   On the Number of Minimal Feedback Vertex Sets

In this section we use a measure based analysis in order to obtain an upper bound of $1.8638^n$ for the number of maximal induced forests (and thus the number of minimal feedback vertex sets) in a graph $G$ on $n$ vertices. It follows from the result of Schwikowski and Speckenmeyer [SS02] that all maximal induced forests and all minimal feedback vertex sets can be enumerated in time $\mathcal{O}(1.8638^n)$.

We also give a lower bound, namely we exhibit an infinite family of graphs, all having $105^{n/10} \approx 1.5926^n$ maximal induced forests. Thus, the worst case running time of the algorithm in [SS02] is between $\Omega(1.5926^n)$ and $\mathcal{O}(1.8638^n)$.

First, we prove the upper bound for the number of maximal induced forests.

**Theorem 3.7.** *A graph $G$ on $n$ vertices contains at most $1.8638^n$ maximal induced forests.*

*Proof.* To prove the theorem, we show that $|\mathcal{M}_G(\emptyset)| \leq 1.8638^n$. We will prove a slightly stronger statement, namely that for any acyclic subset $F$ of $G = (V, E)$, $|\mathcal{M}_G(F)| \leq 1.8638^{n-|F|}$. By Lemma 3.1 we may assume that $F$ is independent. For a graph $G$, an independent set $F$ and a vertex $t \in F$ (we call such a vertex $t$ an *active vertex*), we use the same kind of measure as in the previous section:

$$\mu(G, F, t) := \alpha \cdot |N(t)| + \beta \cdot |V \setminus (F \cup N(t))|,$$

with $\alpha := 0.58$ and $\beta := 0.89823$. In the case where $F = \emptyset$, we set

$$\mu(G, \emptyset) := \beta \cdot |V|.$$

Note, that $\mu(G, F, t) \leq \mu(G, \emptyset) = \beta n$ for every $F$ and $t \in F$. Let $f(G, F) = |\mathcal{M}_G(F)|$ be the number of maximal induced forests containing $F$ and let $f(\mu)$ be a maximum $f(G, F)$ among all triples $(G, F, t)$ and couples $(G, \emptyset)$ of measure at most $\mu$. We claim that

$$f(\mu) \leq 2^\mu.$$

Since for $F = \emptyset$ every vertex of $G$ has weight $\beta$, the claim implies that $|\mathcal{M}_G(\emptyset)| \leq 2^{\beta n} \leq 2^{0.89823n} \leq 1.8638^n$, which proves the theorem.

Let us observe that the claim is true for $\mu = 0$. In fact, for $\mu = 0$ we have that $F = V$. Thus $\mathcal{M}_G(F) = \{V\}$ and $f(0) = 1$. To prove the claim we proceed by induction assuming that $f(\kappa) \leq 2^\kappa$ for every $\kappa < \mu$. Let $(G, F, t)$ be an instance of measure $\mu$.

We consider several cases. As in the previous section, we assume that inside a given case, the hypotheses of all previous cases are assumed to be false.

**Case 1:** $G$ *is not connected.* Denote by $G_1, G_2, \ldots, G_k$ the connected components of $G$. Let $F_i$ denote the intersection of $F$ and the vertices of $G_i$, for $i = 1, 2, \ldots, k$. If the vertices of $V \setminus F$ are present in at least two components, then for all $i \in \{1, \ldots, k\}$, $\mu(G_i, F_i) < \mu(G, F)$ and by the induction assumption,

$$f(\mu) = \prod_{i=1}^k f(G_i, F_i) \leq \prod_{i=1}^k 2^{\mu(G_i, F_i)} = 2^{\sum_{i=1}^k \mu(G_i, F_i)} = 2^\mu.$$

Otherwise, each component which does not contain vertices of $V \setminus F$ has exactly one maximal induced forest (see the next case) and the component including all the vertices of $V \setminus F$ (which determines the overall number of the maximal induced forests) has less vertices than $G$. Hence we may consider that we prove the theorem by two-dimensional induction, the first dimension is the induction on $\mu$, the second dimension is induction on the number of vertices of the underlying graph. The considered case follows from the induction assumption of the second dimension. In fact, this is the only place in the proof where the second dimension is used.

**Case 2:** $F = \emptyset$. If $\Delta(G) \leq 1$ then $\mathcal{M}_G(F) = \{V\}$, that is $f(G, F) = 1$. Otherwise, let $t$ be a vertex of $G$ of degree at least 2. Then every maximal forest either contains $t$, or does not. Thus the number of maximal forests in $G$ is equal to the number of maximal forests containing

$t$, that is $f(G, \{t\})$, plus the number of maximal forests not containing $t$, that is $f(G \setminus \{t\}, \emptyset)$. Since

$$\mu(G, \{t\}, t) \leq \mu - \beta - 2(\beta - \alpha)$$

and

$$\mu(G \setminus \{t\}, \emptyset) \leq \mu - \beta,$$

we use the induction assumption and arrive at

$$f(\mu) \leq f(\mu - \beta - 2(\beta - \alpha)) + f(\mu - \beta) \leq 2^{\mu - \beta - 2(\beta - \alpha)} + 2^{\mu - \beta} \leq 2^\mu.$$

From now on we denote by $t \in F$ an active vertex (if $F \neq \emptyset$ contains no such vertex, we may always choose an arbitrary vertex as active, reducing the measure).

**Case 3:** $V \setminus F = N(t)$. Then by Lemma 3.3, $f(\mu)$ is equal to the number of maximal independent sets in the graph $H$ from Lemma 3.3. Since all vertices of $V \setminus F$ have weight $\alpha$, $H$ has $\mu/\alpha$ vertices. By Theorem 3.5,

$$f(\mu) \leq 3^{\mu/3\alpha} \leq 2^\mu,$$

as $(\log_2 3)/(3\alpha) \leq 1$.

**Case 4:** There is a vertex $v \in N(t)$ such that $gd(v) = 0$. In this case every $X \in \mathcal{M}_G(F)$ contains $v$ and thus $f(G, F) = f(G, F \cup \{v\})$. Since $\mu(G, F \cup \{v\}, t) < \mu$, we have that $f(\mu) \leq 2^\mu$.

Now we assume that $V \setminus F \neq N(t)$, that $F \neq \emptyset$ and that $G$ is connected. Then there is a vertex $v \in N(t)$ such that at least one of its generalized neighbors lies not in $N(t)$ (and thus contributes weight $\beta$ to the measure). Among all such vertices we choose a vertex $v$ of minimum generalized degree. Similarly to the proof of Theorem 3.6, it follows from Lemmata 3.1 and 3.2 that every $X \in \mathcal{M}_G(F)$ must contain either $v$ or at least one of its generalized neighbors.

**Case 5:** $gd(v) = 1$. Every forest $X \in \mathcal{M}_G(F)$ either contains $v$, or does not contain $v$ and contains its generalized neighbor $w_1$. The measure $\mu(G, F \cup \{v\}, t)$ is at most $\mu - \beta$ as $w_1 \notin N(t)$, and the measure $\mu(G \setminus \{v\}, F \cup \{w_1\}, t)$ is at most $\mu - \alpha - \beta$. Hence

$$f(\mu) \leq f(\mu - \beta) + f(\mu - \alpha - \beta) \leq 2^{\mu - \beta} + 2^{\mu - \alpha - \beta} \leq 2^\mu.$$

**Case 6:** $gd(v) = 2$. Let us denote the generalized neighbors of $v$ by $w_1$ and $w_2$ and let us assume that $w_1 \notin N(t)$. Then every forest $X$ from $\mathcal{M}_G(F)$

— Either contains $v$;

— or does not contain $v$ and contains $w_1$;

— or does not contain $v$ and $w_1$ but contains $w_2$.

Let us note that if $w_2 \in N(t)$ and $v$ belongs to a maximal induced forest $X$, then $w_2$ does not belong to $X$. Thus if $w_2 \in N(t)$, then the number of forests in $\mathcal{M}(F)$ is at most

$$f(G \setminus \{w_2\}, F \cup \{v\}) + f(G \setminus \{v\}, F \cup \{w_1\}) + f(G \setminus \{v, w_1\}, F \cup \{w_2\}).$$

Thus

$$\begin{aligned}
f(\mu) &\leq f(\mu - 2\alpha - (\beta - \alpha)) + f(\mu - \alpha - \beta) + f(\mu - 2\alpha - \beta) \\
&\leq 2 \cdot 2^{\mu - \alpha - \beta} + 2^{\mu - 2\alpha - \beta} \leq 2^{\mu}.
\end{aligned}$$

If $w_2 \notin N(t)$, then

$$\begin{aligned}
f(\mu) &\leq f(\mu - \alpha - 2(\beta - \alpha)) + f(\mu - \alpha - \beta) + f(\mu - \alpha - 2\beta) \\
&\leq 2^{\mu + \alpha - 2\beta} + 2^{\mu - \alpha - \beta} + 2^{\mu - \alpha - 2\beta} \leq 2^{\mu}.
\end{aligned}$$

**Case 7:** $gd(v) = 3$. Denote the generalized neighbors of $v$ by $w_1, w_2$, and $w_3$ according to the rule that $w_j \notin N(t)$ and $w_k \in N(t)$ imply $j < k$. Then for every forest $X$ from $\mathcal{M}_G(F)$ holds one of the following

— $X$ contains $v$;

— $X$ does not contain $v$ and contains $w_1$;

— $X$ does not contain $v$ and $w_1$ but contains $w_2$; or

— $X$ does not contain $v$, $w_1$ and $w_2$ but contains $w_3$.

Let $i$ be the number of generalized neighbors of $v$ that are not adjacent to $t$. For $i = 1$, we have

$$\begin{aligned}
f(\mu) &\leq f(\mu - \alpha - (\beta - \alpha) - 2\alpha) + f(\mu - \alpha - \beta) + f(\mu - 2\alpha - \beta) \\
&+ f(\mu - 3\alpha - \beta) \leq 2^{\mu - 2\alpha - \beta} + 2^{\mu - \alpha - \beta} + 2^{\mu - 2\alpha - \beta} + 2^{\mu - 3\alpha - \beta} \leq 2^{\mu}.
\end{aligned}$$

For $i = 2$,

$$\begin{aligned}
f(\mu) &\leq f(\mu - \alpha - 2(\beta - \alpha) - \alpha) + f(\mu - \alpha - \beta) + f(\mu - \alpha - 2\beta) \\
&+ f(\mu - 2\alpha - 2\beta) \leq 2^{\mu - 2\beta} + 2^{\mu - \alpha - \beta} + 2^{\mu - \alpha - 2\beta} + 2^{\mu - 2\alpha - 2\beta} \leq 2^{\mu}.
\end{aligned}$$

For $i = 3$,

$$\begin{aligned}
f(\mu) &\leq f(\mu - \alpha - 3(\beta - \alpha)) + f(\mu - \alpha - \beta) + f(\mu - \alpha - 2\beta) + f(\mu - \alpha - 3\beta) \\
&\leq 2^{\mu + 2\alpha - 3\beta} + 2^{\mu - \alpha - \beta} + 2^{\mu - \alpha - 2\beta} + 2^{\mu - \alpha - 3\beta} \leq 2^{\mu}.
\end{aligned}$$

**Case 8:** $gd(v) \geq 4$. Then every forest $X$ from $\mathcal{M}_G(F)$ either contains $v$ or does not. Thus

$$f(\mu) \leq f(\mu - \alpha - 4(\beta - \alpha)) + f(\mu - \beta) \leq 2^{\mu + 3\alpha - 4\beta} + 2^{\mu - \beta} \leq 2^{\mu}.$$

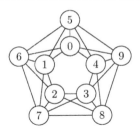

Figure 3.1: Generating graph $C_5 \boxtimes P_2$ used to lower bound the number of maximal induced forests in a graph

$\square$

*Remark* 4. The two tight constraints here are in the case Main 7, when $i = 1$ and when $i = 3$. Again, an improvement of this case would provide a better bound on the number of minimal feedback vertex sets.

Now, we prove the lower bound for the number of maximal induced forests.

**Theorem 3.8.** *There exists an infinite family of graphs all having $105^{n/10} \approx 1.5926^n$ maximal induced forests.*

*Proof.* The infinite family consists of disjoint copies of the graph given in Figure 3.1 (the strong product of a $C_5$ and a $P_2$). The same family of graphs has been used in [BMS05] to show that the number of maximal induced bipartite subgraphs is lower bounded by $1.5926^n$.

A *pair* of vertices in the graph of Figure 3.1 are two vertices whose labels differ by 5. This graph has $5 \cdot 2^4 = 80$ maximal induced forests containing one vertex from 4 of the pairs, $5 \cdot 2^2 = 20$ containing one pair and one vertex from each of the "opposite" pairs and 5 containing two pairs. In total, it has 105 maximal induced forests.

It is clear that maximal induced forests of a disconnected graph are unions of maximal induced forests of its connected components. Their number thus equals the product of the number of maximal induced forests of each component. By taking multiple copies of the graph in Figure 3.1, we get the lower bound of $105^{n/10}$. $\square$

## 3.6   Conclusion

In this chapter we presented an $\mathcal{O}(1.7548^n)$ time algorithm finding a minimum feedback vertex set in an undirected graph on $n$ vertices. We also proved that a graph on $n$ vertices can contain at most $1.8638^n$ minimal feedback vertex sets and that there exist graphs having $105^{n/10} \approx 1.5926^n$ minimal feedback vertex sets. The design and analysis of algorithms establishing the first two results is based on the following three ideas. The first one is considering the complementary problem of maximum induced forest instead the straightforward computing of the feedback vertex set. The second idea is a generalization of the maximum induced problems according to which a subset of vertices $F$ of the given graph $G$ is introduced and the task is to find the largest forest including $F$ as a subset. The third idea is a good choice of the measure of the subproblems recursively generated by the algorithm. This good choice led us to a significantly better worst case running time analysis of the proposed algorithm.

There are a few possible directions of further research related to the topic of this chapter. The first is the design of a faster algorithm for computing a minimum feedback vertex set (or maximum induced forest). Another possible research direction is to ask the same questions as addressed in this chapter to other classes of graphs than forests: Find a maximum induced subgraph that belongs to a certain class $\mathcal{C}$ of graphs. In the literature, there exist algorithms faster than $\mathcal{O}(2^n)$ if $\mathcal{C}$ is, for example,

- the class of $k$-colorable graphs for $k \leq 3$ (see [TT77, Rob86, Jia86, Bei99, FGK09b] for $k = 1$, [Bys04b, RSS07, AT06] for $k = 2$ and [AT06] for $k = 3$),

- the class of graphs of treewidth at most $t$, for $t = o(n/\log n)$ [FV10],

- the class of cluster graphs (see Chapter 9), and

- the class of $d$-regular graphs (see [GRS06]).

It is also easy to give an $\mathcal{O}^*(3^{n/3})$ algorithm for the case where $\mathcal{C}$ is the class of paths and algorithms faster than $\mathcal{O}(2^n)$ follow by a reduction to MINIMUM $k$-HITTING SET for all classes of graphs with a finite number of finite forbidden subgraphs[1], for example split graphs [FH77], cographs, line graphs [Bei70] and trivially perfect graphs [Gol78]. It remains open to find an algorithm faster than $\mathcal{O}(2^n)$ for the cases where $\mathcal{C}$ is the class of chordal graphs, planar graphs, or even outerplanar graphs, for example.

---

[1] see [Wah04, Wah07, FGK+08] and Chapter 9 for algorithms for MINIMUM $k$-HITTING SET

# Chapter 4

# On Bicliques in Graphs

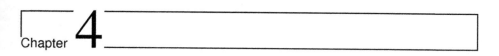

Thoughts without content are empty, intuitions without concepts are blind.

*Immanuel Kant*

Bicliques of graphs have been studied extensively, partially motivated by the large number of applications. One of the main algorithmic interests is in designing algorithms to enumerate all maximal bicliques of a (bipartite) graph. Polynomial time reductions have been used explicitly or implicitly to design polynomial delay algorithms to enumerate all maximal bicliques.

Based on polynomial time Turing reductions, various algorithmic problems on (maximal) bicliques can be studied by considering the related problem for (maximal) independent sets. In this line of research, we show that the maximum number of maximal bicliques in a graph on $n$ vertices is exactly $3^{n/3}$ (up to a polynomial factor). We also give algorithms for various problems related to bicliques, and mainly $\mathcal{O}(1.3642^n)$ time algorithms to compute the number of maximal independent sets and maximal bicliques in a graph.

## 4.1 Introduction

**Bicliques.** Let the vertex sets $X$ and $Y$ be independent sets of a graph $G = (V, E)$ such that $xy \in E$ for all $x \in X$ and all $y \in Y$. The subgraph of $G$ induced by $X \cup Y$ is called a *biclique* of $G$. Furthermore depending on the context and the application area, one also calls the pair $(X, Y)$ or the vertex set $X \cup Y$ a biclique. From a graph–theoretic point of view it is natural to consider a biclique of a graph $G$ as a complete bipartite induced subgraph of $G$. For technical reasons, we prefer to consider a biclique $B \subseteq V$ of a graph $G = (V, E)$ as a vertex set inducing a complete bipartite subgraph of $G$.

**Maximal bicliques.** A biclique $B \subseteq V$ of $G$ is a *maximal biclique* of $G$ if $B$ is not properly contained in another biclique of $G$.

A lot of the research on maximal bicliques and in particular on algorithms to enumerate all maximal bicliques of (bipartite) graphs with polynomial delay is motivated by the various applications of bicliques in (bipartite) graphs. Applications of bicliques in automata and language theory, graph compression, artificial intelligence and biology are discussed in [AVJ98]. An important application in data mining is based on the formal concept analysis [GW96] where each concept is a maximal biclique of a bipartite graph.

**Previous work.** The complexity of algorithmic problems on bicliques has been studied extensively. First results were mentioned by Garey and Johnson [GJ79], among them the $\mathcal{NP}$-completeness of the balanced complete bipartite subgraph problem. The maximum biclique problem is polynomial for bipartite graphs [DKST01], and $\mathcal{NP}$-hard for general graphs [Yan78]. The maximum edge biclique problem was shown to be $\mathcal{NP}$-hard by Peeters [Pee03].

Hochbaum [Hoc98] gives approximation algorithms for node and edge deletion biclique problems. Enumerating maximal bicliques has attracted a lot of attention in the last decade. The algorithms in [NR99, NR02] enumerate all maximal bicliques of a bipartite graph as concepts during the construction of the concept lattice. Nowadays there are polynomial delay enumeration algorithms for maximal bicliques in bipartite graphs [DdFS07, MU04] and general graphs [DdFS05, MU04]. There are also polynomial delay algorithms to enumerate all maximal non–induced bicliques of a graph [AAC+04, DdFS07][1] and exponential time algorithms to find non–induced bicliques of a given size [FGK+09a].

Prisner studied various aspects of bicliques in graphs. Among others, he showed that the maximum number of maximal bicliques in a bipartite graph on $n$ vertices is $2^{n/2}$. He established a lower bound of $3^{n/3}$ and an upper bound of $1.6181^n$ (up to a polynomial factor) on the maximum number of maximal bicliques in a graph on $n$ vertices [Pri00].

**Our Results.** We use a simple polynomial time Turing reduction to transform results on maximal independent sets into results on maximal bicliques. We also improve upon Prisner's upper bound and give a simple proof that the maximum number of maximal bicliques in a graph on $n$ vertices is at most $n \cdot 3^{n/3}$. On the algorithmic side, our main result is an $\mathcal{O}(1.3642^n)$ time algorithm to count all maximal independent sets in a graph, which is established by using a measure based analysis as described in Chapter 2. We show how to use it to count all maximal bicliques of a graph within the same time bound and also provide a lower bound for the running time of this algorithm.

## 4.2 Polynomial Time Reductions

There is a natural relation between independent sets (and cliques) on one hand and bicliques on the other hand. Thus it is not surprising that polynomial–time Turing reductions (in fact mainly Karp reductions) have been used in various hardness proofs for problems on bicliques [GJ79]. The famous polynomial delay algorithm of Johnson and Papadimitriou to enumerate all maximal independent sets [JYP88] is used explicitly or implicitly in polynomial delay algorithms to enumerate maximal (non–induced) bicliques in (bipartite) graphs [AAC+04, DdFS05, DdFS07].

---

[1]When the condition that $X$ and $Y$ are independent sets in the definition of a biclique is omitted, then $(X, Y)$ is called a *non–induced biclique* of $G$. In this case a different maximality notion is used. See for example [AAC+04].

The first reduction simply recalls an often used argument.

**Lemma 4.1** (Property A). *Let $G = (V, E)$ be a bipartite graph. Let $H$ be the bipartite complement of $G$. Then $B$ is a (maximal) biclique of $G$ if and only if $B$ is a (maximal) independent set of $H$.*

The above lemma implies, among others, that any algorithm enumerating all maximal independent sets within delay $f(n)$ can be transformed into an algorithm enumerating all maximal bicliques of a bipartite graph within delay $f(n)$. The known tight bound of $2^{n/2}$ for the maximum number of maximal bicliques in a bipartite graph given in [Pri00] follows easily from Property A and the corresponding bound for maximal independent sets in [HT93]. Based on this property, Yannakakis observed that the problem of finding a maximum biclique in a bipartite graph can be solved in polynomial time [Yan78].

The following property is central for this chapter.

**Lemma 4.2** (Property B). *Let $G = (V, E)$ be a graph. For every $v \in V$, the graph $H_v$ is the graph with vertex set $V(H_v) := N(v) \cup N^2(v)$. Its edge set $E(H_v)$ consists of the following edges:*

- *$xy \in E(H_v)$ if $xy \in E$ and $x, y \in N(v)$,*

- *$xy \in E(H_v)$ if $xy \in E$ and $x, y \in N^2(v)$,*

- *$xy \in E(H_v)$ if $xy \notin E$, $x \in N(v)$ and $y \in N^2(v)$.*

*Then $B \subseteq V$ is a (maximal) biclique of $G$ if and only if $B \setminus \{v\}$ is a (maximal) independent set of a graph $H_v$ for some $v \in B$.*

*Proof.* Let $B$ be a (maximal) biclique of $G$. Take some $v \in B$. Then $B \subseteq \{v\} \cup N(v) \cup N^2(v)$ in $G$, where the independent sets $X$ and $Y$ of the biclique $B$ satisfy $X \subseteq N(v)$ and $Y \subseteq \{v\} \cup N^2(v)$. Since $B$ is a biclique and by the construction of $H$, we obtain that $B \setminus \{v\}$ is an independent set. On the other hand, if $B'$ is a (maximal) independent set of $H_v$, for some $v \in V$, then $B' \cap N(v)$ is an independent set of $G[N(v)]$ and $B' \cap N^2(v)$ is an independent set of $G[N^2(v)]$. Hence $B'$ is a biclique of $G \setminus v$ and $B' \cup \{v\}$ is a biclique of $G$.

Finally, due to the correspondence between bicliques and independent sets, this also holds for maximality by inclusion of vertices. □

The corresponding Turing reduction does not increase the number of vertices, since $|V(H_v)| \le |V| - 1$. Thus this reduction is useful for exponential time algorithms.

**Corollary 4.3.** *Given an algorithm to find a maximum independent set (respectively to count all independent sets of size $k$) of a graph in time $\mathcal{O}^*(c^n)$, there exists an algorithm to find a maximum biclique (respectively to count all bicliques of size $k$) of a graph in time $\mathcal{O}^*(c^n)$.*

*Proof.* To find a maximum biclique of a graph $G = (V, E)$, compute a maximum independent set for each $H_v$, $v \in V$, constructed according to Property B and return the largest set of vertices found. To count all bicliques of size $k$ of a graph $G = (V, E)$ on $n$ vertices, order the vertices of $G$: $V := \{v_1, v_2, \ldots, v_n\}$. For $i = 1, \ldots, n$, compute the number of independent sets of size $k - 1$ of $H_{v_i}^i$ where $H_{v_i}^i$ is obtained from $G^i = G \setminus \{v_1, v_2, \ldots, v_{i-1}\}$ using Property B. Adding up the results gives the number of bicliques of size $k$ of $G$. $\qquad \square$

By this corollary and the algorithms in [Rob86, Wah08], a maximum biclique of a graph can be found in time $\mathcal{O}(1.2109^n)$ and all maximum bicliques of a graph can be counted in time $\mathcal{O}(1.2377^n)$.

Note that Corollary 4.3 is not directly applicable to use an algorithm for counting maximal independent sets to count the maximal bicliques of a graph. The issues are that double–counting has to be avoided at the same time as the maximality of each counted biclique has to be ensured.

## 4.3 Combinatorial Bound for the Number of Maximal Bicliques

The maximum number of maximal bicliques in a graph on $n$ vertices has been studied by Prisner [Pri00]. He settled the question for bipartite graphs. The maximum number of maximal bicliques in a bipartite graph on $n$ vertices is precisely $2^{n/2}$. For general graphs the question remained open. He established a lower bound of $3^{n/3}$ and an upper bound of $(1.618034^n + o(1)) \cdot n^{5/2}$ for the maximum number of maximal bicliques in a graph on $n$ vertices. We settle the question via an elegant proof based on Property B.

**Theorem 4.4.** *The maximum number of maximal bicliques in a graph is at most $n \cdot 3^{n/3}$.*

*Proof.* Let $n$ be a positive integer and let $G$ be any graph on $n$ vertices. Applying Property B, for every vertex $v \in V$, there is a one–to–one correspondence between the maximal bicliques $B$ of $G$ satisfying $v \in B$ and the maximal independent sets $B - v$ of the graph $H_v$. By Theorem 3.5 on page 52, the maximum number of maximal independent sets in a graph on $n$ vertices is $3^{n/3}$. Thus the number of maximal bicliques containing vertex $v$ is at most $3^{n/3}$ for each $v \in V$. Consequently $G$ has at most $n \cdot 3^{n/3}$ maximal bicliques. $\qquad \square$

**Corollary 4.5.** *The maximum number of maximal bicliques in a graph is $3^{n/3}$ (up to a polynomial factor).*

## 4.4 Counting Algorithms

A problem related to enumerating all maximal bicliques of a graph is to compute the number of maximal bicliques of a graph faster than by simply enumerating all of them. By property B, an algorithm to count all maximal independent sets of a graph could be a cornerstone to design such an algorithm. However no non–trivial algorithm for counting maximal independent sets was known prior to our work. It is known that the counting problem for maximal independent sets is #$\mathcal{P}$–complete even when restricted to chordal graphs [OUU08]. Hence our goal is to construct a fast exponential time algorithm solving this problem.

## 4.4.1    Algorithm to Count all Maximal Independent Sets

We would first like to say a word of precaution. Even if the problems of counting all maximal independent sets of a graph seems very similar to the problem of counting all maximum independent sets of a graph, or all independent sets of a given size $k$, there is a fundamental difference coming from the notion of maximality. All fast algorithms for counting all independent sets of maximum size or of size $k$ [DJ02, DJW05, FK05, Wah08] rely on a branching strategy similar to the one of Algorithm **mis** in Chapter 2: vertices that are decided not to be in the counted independent sets of a branch can be deleted and removed from further consideration, and graphs of maximum degree 2 can be handled in polynomial time. But if the algorithm is supposed to count all maximal independent sets, this strategy does not work (unless $\mathcal{P} = \#\mathcal{P}$). Consider a graph $G = (F \cup M, E)$ for which we would like to count all maximal independent sets of $G$ that are included in $F$. In other words, $M$ is the set of vertices that have been decided not to be in any maximal independent set in the current branch, but for each of them, a neighbor must be added to ensure the maximality of the counted independent sets. By a simple reduction from $\#\text{SAT}$, it can be shown that this problem is $\#\mathcal{P}$-hard even when $G[F]$ has maximum degree 1 (an edge in $G[F]$ corresponds to a variable, its end points to the **true**/**false** value of this variable, and the vertices in $M$ correspond to the clauses of the formula).

Let $G = (F, M, E)$ be a *marked graph* which are the graphs dealt with by our algorithm. Vertices of $F$ are called *free* and vertices of $M$ are called *marked*. Let $u$ be a vertex of $F \cup M$. The degree of $u$ is the number of neighbors in $F \cup M$ and is denoted by $d(u)$. Given a set $D \subseteq (F \cup M)$, the set $N(u) \cap D$ is denoted by $N_D(u)$ and its cardinality is denoted by $d_D(u)$. For a marked graph $G = (F, M, E)$, the marked graph induced by the vertex sets $F' \subseteq F$ and $M' \subseteq M$ is $G[F', M'] = (F', M', E \cap ((F' \cup M') \times (F' \cup M')))$.

The following notions are crucial for our algorithm. A set $S \subseteq F$ is a maximal independent set of a marked graph $G = (F, M, E)$ if $S$ is a maximal independent set of $G[F]$. We say that the maximal independent set $S$ of $G$ satisfies property $\Pi$ if each vertex of $M$ has a neighbor in $S$.

Given a marked graph $G$, our algorithm computes the number of maximal independent sets of $G = (F, M, E)$ satisfying $\Pi$. Thus, a marked vertex $u$ is used to force that each maximal independent set $S$ of $G$ counted by the algorithm contains at least one free neighbor of $u$. This is particularly useful to guarantee that only maximal independent sets of the input graph are counted. In the remainder of this section, we suppose that $G$ is a connected graph, otherwise the algorithm is called for each of its connected components, and the product of the results gives the number of maximal independent sets of $G$ satisfying $\Pi$.

Given a simple graph $G' = (V, E)$, #MaximalIS$(G = (V, \emptyset, E))$ returns the number of maximal independent sets of $G'$. See Figure 4.1 for the description of the algorithm.

We emphasize that all the halting ((H1)–(H2)) and simplification ((S1)–(S7)) rules are necessary for our running time analysis in Subsections 4.4.3 and 4.4.4. The branching rule (B) selects a vertex $u$, orders its free neighbors in a list $BL(u) = [v_1, v_2, \ldots, v_{d_F(u)}]$ and makes a recursive call (that is a branching) counting all maximal independent sets containing $u$, and a recursive call for each $i = 1, 2, \ldots, d_F(u)$ where it counts all maximal independent sets containing $v_i$ but none of $v_1, v_2, \ldots, v_{i-1}$.

---

**Algorithm** #MaximalIS$\big(G = (F, M, E)\big)$
**Input:** A marked graph $G = (F, M, E)$.
**Output:** The number of maximal independent sets of $G$ satisfying $\Pi$.
`// Simplification rules`
**if** $F \cup M$ *is empty* **then**
    | **return** 1                                                     (H1)

**if** *there exists* $u \in M$ *such that* $d_F(u) = 0$ **then**
    | **return** 0                                                     (H2)

**if** *there exists* $u \in M$ *such that* $N_F(u) = \{v\}$ **then**
    | **return** #MaximalIS$\big(G[F \setminus N[v], M \setminus N(v)]\big)$         (S1)

**if** *there exists* $u \in F$ *such that* $d_F(u) = 0$ **then**
    | **return** #MaximalIS$\big(G[F \setminus N[u], M \setminus N(u)]\big)$         (S2)

**if** *there exists* $u, v \in M$ *such that* $\{u, v\} \in E$ **then**
    | **return** #MaximalIS$\big((F, M, E \setminus \{u, v\})\big)$             (S3)

**if** *there exists* $u, v \in F$ *such that* $N[u] = N[v]$ **then**
    | $count \leftarrow$ #MaximalIS$\big(G[F \setminus \{v\}, M]\big)$
    | Let $\text{MIS}_u$ be the number of maximal independent sets computed by
    | #MaximalIS$\big(G[F \setminus \{v\}, M]\big)$ containing $u$
    | **return** $\text{MIS}_u + count$                                 (S4)

**if** *there exists* $u \in M$ *and* $v \in N(u)$ *such that* $N[v] \subseteq N[u]$ **then**
    | **return** #MaximalIS$\big(G[F, M \setminus \{u\}]\big)$              (S5)

**if** *there exists* $u, v \in M$ *such that* $N(u) = N(v)$ **then**
    | **return** #MaximalIS$\big(G[F, M \setminus \{v\}]\big)$              (S6)

**if** *there exists* $u \in F \cup M$ *and* $v \in F$ *such that* $N(u) = N(v)$ **then**
    | **return** #MaximalIS$\big(G[F \setminus \{v\}, M]\big)$             (S7)

`// Branching rule`                                                         (B)
**if** *there exists a marked vertex* $u$ *with* $d(u) = 2$ **then**
    L Choose $u$
**else**
    | Choose a vertex $u \in (F \cup M)$ such that
         (i) $u$ has minimum degree among all vertices in $F \cup M$
         (ii) among all vertices fulfilling (i), $u$ has a neighbor of maximum degree
         (iii) among all vertices fulfilling (i) and (ii), $u$ has maximum dual degree
Let $BL(u) \leftarrow [v_1, \ldots, v_{d_F(u)}]$ be an ordered list of $N_F(u)$ such that:
    (i) $v_1$ is a vertex of $N_F(u)$ having a minimum number of neighbors in $V \setminus N(u)$; if there are
    several choices, choose $v_1$ of minimum degree
    (ii) append (in any order) the vertices of $N(v_1) \cap N_F(u)$ to the ordered list
    (iii) append $N_F(u) \setminus N[v_1]$ ordered by increasing number of neighbors in $V \setminus N(u)$
$count \leftarrow 0$
**if** $u$ *is free* **then**          `// select u (to be in the current maximal independent set)`
    | $count \leftarrow$ #MaximalIS$\big(G[F \setminus N[u], M \setminus N(u)]\big)$
**foreach** $v_i \in BL(u)$ **do**          `// mark each vertex of M' and select vi`
    | $M' \leftarrow \{v_j \in \text{BL}(u) : 1 \leq j < i \text{ and } \{v_j, v_i\} \notin E\}$
    | $count \leftarrow count +$ #MaximalIS$\big(G[F \setminus (M' \cup N[v_i]), (M \cup M') \setminus N(v_i)]\big)$
**return** $count$

Figure 4.1: Algorithm #MaximalIS counting all maximal independent sets

---

The selected vertex $u$ is chosen according to three criteria (i)–(iii). By (i), $u$ has minimum degree, which ensures either that the algorithm makes few recursive calls or that many vertices are removed in each branching. By (ii), $u$ has a neighbor of maximum degree among all vertices satisfying (i). If the degree of this neighbor is high, then many vertices are removed in at least one recursive call. If the degree of this vertex is low, every vertex of minimum degree has no high–degree neighbor. This property is exploited in the analysis of our algorithm, which considers a decrease in the degree of a vertex of small degree more advantageous than a decrease in the degree of a high–degree vertex. Similarly, (iii) ensures either many recursive calls where many vertices are removed or a knowledge on the degrees of the neighbors of a vertex of minimum degree. The ordered list $\mathrm{BL}(u)$ is defined in this way to ensure that for certain configurations of $N^2[u]$, simplification rule (S1) or a (fast) subsequent branching on a marked vertex of degree 2 is applied in many recursive calls.

## 4.4.2   Correctness of #MaximalIS

We show the correctness of the branching and simplification rules of #MaximalIS. **(H1)** If the input graph has no vertices then the only maximal independent set is the empty set. **(H2)** If there is a marked vertex $u$ without any free neighbor then there is no maximal independent set satisfying $\Pi$. **(S1)** If a marked vertex $u$ has only one free neighbor, it has to be in any maximal independent set to satisfy $\Pi$. **(S2)** By maximality, each free vertex without any free neighbor has to belong to all maximal independent sets. **(S3)** Since marked vertices cannot belong to any maximal independent set, edges between two marked vertices are irrelevant and can be removed. **(S4)** Suppose $u, v \in F$ are two free vertices and $N[u] = N[v]$. Every maximal independent set containing a neighbor of $u$ does not contain $v$. Moreover, every maximal independent set containing $u$ can be replaced by one containing $v$ instead of $u$. Thus, it is sufficient to remove $v$ and to return the number of maximal independent sets containing a neighbor of $u$ plus twice the number of maximal independent sets containing $u$. (Note that the algorithm can easily be implemented such that the number of maximal independent sets containing $u$ is obtained from the recursive call. For example, keep a counter to associate to each free vertex the number of maximal independent sets containing this vertex.) **(S5)** If $u \in M$ has a neighbor $v$ such that all neighbors of $v$ are also neighbors of $u$, then every maximal independent set of $G \setminus u$ must contain a vertex of $N[v] \setminus \{u\}$ and thus a neighbor of $u$ in $G$. **(S6)** If two marked vertices have the same neighborhood then one of them is irrelevant. **(S7)** Let $v$ be a free vertex and $u$ a vertex such that $N(u) = N(v)$, and thus $u$ and $v$ are non adjacent. Hence every maximal independent set containing a neighbor of $u$ does not contain $v$ and every maximal independent set containing $u$ (if $u$ is free) also contains $v$. Thus the number of maximal independent sets is the same as for $G \setminus v$.

**(B)** The algorithm considers the two possibilities that either $u$ or at least one neighbor of $u$ is in the current maximal independent set. By induction and the fact that $N[u]$ is removed if the algorithm decides to add $u$ to the current maximal independent set, every maximal independent set containing $u$ is counted and it is counted only once. Consider the possibility that at least one neighbor of $u$ is in the current maximal independent set and let $v_i$ be the first such neighbor in the ordered list $\mathtt{BL}(u)$, containing all the free neighbors of $u$. That no maximal independent set containing a vertex appearing before $v_i$ in $\mathtt{BL}(u)$ is counted, is ensured by either its deletion (because it is a neighbor of $v_i$) or the marking of this vertex. So, every maximal independent set containing $v_i$ but neither $u$ (removed as it is a neighbor of $v_i$) nor a vertex appearing before $v_i$ in $\mathtt{BL}(u)$ is counted exactly once.

### 4.4.3  Running Time Analysis of #MaximalIS

To analyze the running time of our algorithm, we use the following measure $\mu(G)$ of a marked graph $G$.

$$\mu := \mu\big(G = (F, M, E)\big)$$
$$:= \sum_{i=1}^{n-1} w_i |V_i| + \mathsf{K}_\delta(G \text{ has no marked vertex of degree } 2)M_2$$

The weights $M_2$ and $w_i$, $1 \le i \le n-1$ are real non–negative numbers that will be fixed later. For $1 \le i \le n-1$, $V_i$ denotes the set of vertices of degree $i$ in $G$. The following values will be useful in the analysis.

$$\Delta w_i := \begin{cases} w_i - w_{i-1} & \text{if } 2 \le i \le n-1 \\ w_1 & \text{if } i = 1 \end{cases}$$

To further simplify the forthcoming analysis, we assume:

$$w_i = w_{i+1}, \qquad\qquad 4 \le i \le n-1,$$
$$w_{i-1} \le w_i, \qquad\qquad 2 \le i \le n-1, \text{ and}$$
$$\Delta w_i \ge \Delta w_{i+1}, \qquad\qquad 1 \le i \le n-1.$$

It is not hard to see that an application of a simplification rule will not increase $\sum_{i=1}^{n-1} w_i |V_i|$. Furthermore no simplification rule can be applied more than $n$ times, respectively $m$ times for (S3). As in every simplification rule and every branch of the branching rule, at least one vertex or edge is removed, we set $\eta(G) := n + m$ and use Lemma 2.5 on page 37 to upper bound the running time of the algorithm. Let $T(\mu) = 2^\mu$.

We only have to analyze the changes in measure when applying branching rule (B).
**Case 1:** (B) is applied to a marked vertex $u$ with $d(u) = 2$.
Let $v_1$ and $v_2$ be its two neighbors. By (S3), that is since (S3) could not be applied, $v_1, v_2 \in F$, and by (S3), $d(v_1), d(v_2) \ge 2$.

(a) Suppose $d(v_1) = d(v_2) = 2$. For $i \in \{1, 2\}$, let $x_i$ be the other neighbor of $v_i$. If $d(x_1) = d(x_2) = 1$ then the algorithm deals with a component of constant size, and the number of maximal independent sets of such a component can be computed in constant time. Suppose now that $d(x_1) \ge 2$. In the first branch (or subproblem) $u$, $v_1$ and $x_1$ are removed. In the second branch $u$, $v_2$ and $x_2$ are removed. In both branches, the graph might not have a marked vertex of degree 2 any more. Thus, the corresponding constraint is

$$T(\mu) \geq T(\mu - 3w_2 + M_2) + T(\mu - w_1 - 2w_2 + M_2).$$

(b) Suppose $d(v_1) \geq 3$ and $d(v_2) \geq 2$. In the first branch $u$, $v_1$ and at least two other neighbors of $v_1$ are removed. In the second branch $u$, $v_2$ and the other neighbors of $v_2$, at least one, are removed. Thus, the corresponding constraint is $T(\mu) \geq T(\mu - 2w_1 - w_2 - w_3 + M_2) + T(\mu - w_1 - 2w_2 + M_2)$. Since $w_2 \leq w_3$ and $w_2 \leq 2w_1$ (recall that $\Delta w_1 \geq \Delta w_2$), it follows that $3w_2 \leq 2w_1 + w_2 + w_3$ and thus the constraint imposed in case (b) is not stronger than the one of case (a) by the Dominance property on page 41.

**Case 2:** Vertex $u$ is chosen by the *else* statement of (B).
Thus $u$ satisfies the conditions (i), (ii) and (iii). Let $[v_1, \ldots, v_{d_F(u)}]$ be the *Branching List*, short BL($u$), built by the algorithm. Given a vertex $v_i$, $1 \leq i \leq d_F(u)$, of BL($u$), we denote by Op($v_i$) the operation of adding $v_i$ to the current maximal independent set, removing $N[v_i]$ and marking the vertices $v_1, \ldots, v_{i-1}$ that are not adjacent to $v_i$.

Let $\Delta_u$ denote the gain on the measure obtained by adding $u$ to the current maximal independent set. Removing $u$ and its neighbors from the graph decreases $\mu(G)$ by $w_{d(u)} + \sum_{v \in N(u)} w_{d(v)}$. Moreover, the decrease of the degrees of vertices in $N^2(u)$ implies a gain of $\sum_{x \in N^2(u)} (w_{d(x)} - w_{d(x) - d_{N(u)}(x)})$. Let $M_2(u)$ be equal to $M_2$ if the subinstance obtained from adding $u$ to the current maximal independent set has a marked vertex of degree 2 after exhaustively applying all the simplification rules, and equal to 0 otherwise. Then,

$$\Delta_u := w_{d(u)} + \sum_{v \in N(u)} w_{d(v)} + \sum_{x \in N^2(u)} (w_{d(x)} - w_{d(x) - d_{N(u)}(x)}) + M_2(u).$$

Let $\Delta_{\mathsf{Op}(v_i)}$ denote the gain on the measure when $v_i \in$ BL($u$), $1 \leq i \leq d_F(u)$, is selected and added to the maximal independent set. Again, by selecting vertex $v_i$ the vertices of $N[v_i]$ are removed and thus a gain of $w_{d(v_i)} + \sum_{x \in N(v_i)} w_{d(x)}$ is obtained. Since neighbors of vertices of $N^2(v_i)$ have been removed, we gain $\sum_{y \in N^2(v_i)} (w_{d(y)} - w_{d(y) - d_{N(v_i)}(y)})$. The measure further decreases whenever among the marked vertices of $\{v_1, \ldots, v_{i-1}\}$, some of them have only one remaining free neighbor after the deletion of $N[v_i]$. By direct application of simplification rule (S1), these vertices and their neighbors are also removed from the graph. We denote this *extra* gain by $\mathrm{marked}_1(\mathsf{Op}(v_i))$ Thus,

$$\Delta_{\mathsf{Op}(v_i)} := w_{d(v_i)} + \sum_{x \in N(v_i)} w_{d(x)} + \sum_{y \in N^2(v_i)} (w_{d(y)} - w_{d(y) - d_{N(v_i)}(y)})$$
$$+ \mathrm{marked}_1(\mathsf{Op}(v_i)) + M_2(v_i).$$

Putting all together, we obtain the following general constraint for case 2:

$$T(\mu) \geq T(\mu - \Delta_u) + \sum_{v_i \in \mathrm{BL}(u)} T(\mu - \Delta_{\mathsf{Op}(v_i)})$$

Finally, we conclude the time analysis by the measure based method described in Chapter 2. We solve the corresponding convex program and establish an upper bound on the worst case running time of our algorithm. Using the weights $M_2 = 0.2$, $w_1 = 0.37962$, $w_2 = 0.41133$, $w_3 = 0.44244$, and $w_4 = 0.44804$ we obtain:

**Theorem 4.6.** *Algorithm* #MaximalIS *counts all maximal independent sets of a given graph $G$ in time $\mathcal{O}(1.3642^n)$, where $n$ is the number of vertices of $G$.*

For our algorithm analysis the number of constraints is rather moderate and therefore we are able to provide for the interested reader the details of the analysis and list all possible worst cases in the next subsection.

### 4.4.4   Detailed Running Time Analysis of #MaximalIS

In this subsection we outline a detailed running time analysis of Algorithm #MaximalIS. The branching corresponding to the selection of a marked vertex of degree 2 has already been analyzed in detail in our high level analysis in Subsection 4.4.3. Here we give a list of cases, corresponding to the analysis in Case 2 in Subsection 4.4.3. Each case has a number, a condition telling us in which case we are, a picture and a constraint on the measure of an instance based on the measure of the created subinstances in this case. For those cases, where it is not immediate how the constraint is obtained, a comment is added observing facts needed to obtain it.

Denote the neighbors of $u$ by $v_1, v_2, \ldots, v_{d(u)}$. For a selected vertex $u$, we say that $x$ is an *external* neighbor of a vertex $v \in N(u)$ if $x$ is a vertex of $N(v) \setminus N[u]$.

Note that the algorithm can apply the branching rule on an $r$-regular graph, $2 \leq r \leq 4$. However, when dealing with such an $r$-regular graph any subsequent recursive calls will never be on an $r$-regular graph again. Thus, these graphs are not relevant to establish the running time bound (see Subsection 2.8.1). If the graph is 1-regular, then the algorithm would treat it in polynomial time since the size of each connected component is bounded by a constant.

In the following case analysis, cases number 1 (with $d(x_1) = 4$), 18 and 21 correspond to the tight cases.

| 1 | $d(u) = 1, d(v_1) = 2$ |
|---|---|

$v_1 \quad x_1$

$u$ ●━━●━━●⋯

$$T(\mu) \geq T(\mu - w_1 - w_2 - \Delta w_{d(x_1)}) + T(\mu - w_1 - w_2 - w_{d(x_1)})$$

| 2 | $d(u) = 1, d(v_1) \geq 3$ |
|---|---|

$v_1$

$u$ ●━━●⪇

$$T(\mu) \geq T(\mu - w_1 - w_{d(v_1)}) + T(\mu - w_{d(v_1)} - (d(v_1) - 1) \cdot w_1 - w_2)$$

**Comment:** $v_1$ has a neighbor of degree at least 2, otherwise $N[v_1]$ is a connected component.

| 3 | $d(u) = 2, d(v_1) = 2, d(v_2) = 3, d(x_1) = 2, x_1$ being the other neighbor of $v_1$ |
|---|---|

$v_1 \quad x_1$

$u$ ●⟨ $v_2$ ●━━●

$$T(\mu) \geq T(\mu + w_1 - 3w_2 - w_3) + T(\mu - 2w_2 - w_3) + T(\mu - 5w_2 - w_3)$$

**Comment:** $\{v_1, v_2\} \notin E$, as $d(x_1) \neq d(v_2)$. In the branch where $v_2$ is selected, $x_1$ is also selected by (S1) as $v_1$ becomes marked and has a unique neighbor. As $N(u) \neq N(x_1)$, which is ensured by (S6) and (S7), $x_1$ and $v_2$ are not adjacent.

| 4 | $d(u) = 2, d(v_1) = 2, d(v_2) = 3, d(x_1) \geq 3$ |
|---|---|

$$T(\mu) \geq T(\mu - 2w_2 - w_3) + T(\mu - w_2 - 2w_3) + T(\mu - 2w_2 - 4w_3)$$

**Comment:** $\{v_1, v_2\} \notin E$, otherwise $N[u] = N[v_1]$ and (S4) or (S5) would apply. When $v_2$ is selected, $x_1$ is also selected by (S1). By the selection rule of $u$, $d(x_1) = 3$ and no common neighbor of $v_2$ and $x_1$ has degree 2. If $v_2$ and $x_1$ are adjacent, the last branch can be ignored as the instance has no maximal independent set by halting rule (H2). For analyzing the last branch, also note that $w_3 \leq 2w_2$ as $\Delta w_3 \leq \Delta w_2$.

| 5 | $d(u) = 2, d(v_1) = 3, d(v_2) = 3$ |
|---|---|

$$T(\mu) \geq T(\mu - w_2 - 2w_3) + 2T(\mu - 4w_3)$$

**Comment:** The vertices of degree 2 in $N^2(u)$ are not adjacent to both $v_1$ and $v_2$ (otherwise they have the same open neighborhood as $u$). Moreover, two adjacent vertices in $N^2(u)$ of degree 2 are not adjacent to the same vertex in $N(u)$ due to the simplification rules. So, they have neighbors outside $N[u]$ of degree at most 3.

| 6 | $d(u) = 2, d(v_1) = 2, d(v_2) \geq 4$ |
|---|---|

$$T(\mu) \geq T(\mu - 2w_2 - w_4) + T(\mu - 3w_2) + T(\mu - 6w_2 - w_4)$$

**Comment:** $v_1$ and $v_2$ are not adjacent due to (S4) and (S5). If they have a common neighbor, ignore the last branch. In the last branch, $v_2$ and the external neighbor of $v_1$ are selected.

| 7 | $d(u) = 2, d(v_1) \geq 3, d(v_2) \geq 4$ |
|---|---|

$$T(\mu) \geq T(\mu - w_2 - w_3 - w_4) + T(\mu - 3w_2 - w_3) + T(\mu - 4w_2 - w_4)$$

| 8 | $d(u) = 3, d(v_1) = 3, d(v_2) = 3, d(v_3) \geq 5, v_1$ and $v_2$ are adjacent |
|---|---|

$$T(\mu) \geq T(\mu - 3w_3 - w_4) + 2T(\mu - 4w_3) + T(\mu - 9w_3 - w_4)$$

**Comment:** $v_1$ and $v_2$ are not adjacent to $v_3$, otherwise (S4) or (S5) would apply as $v_1$ or $v_2$ would have the same closed neighborhood as $u$. Moreover, $v_1$ and $v_2$ do not share the same external neighbor otherwise $v_1$ and $v_2$ have the same closed neighborhood. If $v_3$ has a common neighbor in $N^2(u)$ with $v_1$ or $v_2$, then ignore the last branch, otherwise $v_3$ and both external neighbors of $v_1$ and $v_2$ are selected.

| 9 | $d(u) = 3, d(v_1) = 3, d(v_2) = 3, d(v_3) \geq 5, N(u)$ is independent, in the last branch $v_1$ and $v_2$ disappear by simplification rules |
|---|---|

$$T(\mu) \geq T(\mu - 3w_3 - w_4) + 2T(\mu + w_2 - 5w_3) + T(\mu - 7w_3 - w_4)$$

**Comment:** In this case, when $v_3$ is selected, $v_1$ and $v_2$ are removed by recursively applying the simplification rules.

| 10 | $d(u) = 3, d(v_1) = 3, d(v_2) = 3, d(v_3) \geq 5, N(u)$ is independent, in the last branch $v_1$ (or $v_2$) does not disappear by simplification rules |
|---|---|

$$T(\mu) \geq T(\mu - 3w_3 - w_4) + 2T(\mu + w_2 - 5w_3) + T(\mu + 2w_2 - 7w_3 - w_4 - M_2)$$

**Comment:** In the last branch $v_1$ and $v_2$ are marked and become of degree 2. Therefore a marked vertex of degree 2 appears $(-M_2)$.

| 11 | $d(u) = 3, d(v_1) = 3, d(v_2) = 3, d(v_3) \geq 5, v_1$ and $v_2$ are not adjacent, $v_3$ is adjacent to $v_1$ and $v_2$ |
|---|---|

$$T(\mu) \geq T(\mu + 2w_2 - 5w_3 - w_4) + T(\mu + w_1 - 4w_3 - w_4) + 2T(\mu - 5w_3 - w_4)$$

**Comment:** The external neighbors of $v_1$ and $v_2$ have degree 3, otherwise $v_1$ or $v_2$ would have a neighbor of higher degree or higher dual degree and would have been selected for branching instead of $u$. Moreover, the external neighbors of $v_1$ and $v_2$ are distinct, otherwise (S6) or (S7) would apply. Finally, note that $\mathrm{BL}(u) = [v_1, v_3, v_2]$ or $\mathrm{BL}(u) = [v_2, v_3, v_1]$. and are distinct.

| 12 | $d(u) = 3, d(v_1) = 3, d(v_2) = 3, d(v_3) \geq 5, v_1$ and $v_2$ are not adjacent, $v_3$ is adjacent to $v_2$ (or $v_1$) |
|---|---|

$$T(\mu) \geq 2T(\mu + w_2 - 4w_3 - w_4) + T(\mu + w_2 - 6w_3 - w_4) + T(\mu - 6w_3)$$

**Comment:** $\mathrm{BL}(u) = [v_2, v_3, v_1]$ and the external neighbor of $v_2$ has degree 3, otherwise $v_2$ would have been selected for branching as it has either a neighbor of higher degree or higher dual degree than $u$.

| 13 | $d(u) = 3, d(v_1) \geq 3, d(v_2) \geq 4, d(v_3) \geq 5$ |
|---|---|

$$T(\mu) \geq T(\mu - 2w_3 - 2w_4) + T(\mu - 4w_3) + T(\mu - 4w_3 - w_4) + T(\mu - 5w_3 - w_4)$$

| 14 | $d(u) = 3, d(v_1) = 3, d(v_2) = 3, d(v_3) = 4, v_1$ and $v_2$ are adjacent |
|---|---|

$$T(\mu) \geq T(\mu - w_3 - 3w_4) + 2T(\mu - 2w_3 - 2w_4) + T(\mu - 8w_3 - w_4)$$

**Comment:** $v_1$ and $v_2$ are not adjacent to $v_3$ because of (S4) and (S5) and they have distinct (by (S4) and (S5)) external neighbors of degree 3 or 4 (by the selection rule of $u$). If $v_3$ has a common neighbor with $v_1$ or $v_2$ (except $u$), ignore the last branch.

| 15 | $d(u) = 3, d(v_1) = 3, d(v_2) = 3, d(v_3) = 4, v_1$ and $v_2$ are not adjacent, $v_3$ is adjacent to $v_1$ and $v_2$ |
|---|---|

$$T(\mu) \geq T(\mu + 2w_2 - 5w_3 - w_4) + T(\mu + w_1 + w_2 - 5w_3 - w_4) + T(\mu + 2w_2 - 6w_3 - w_4) + T(\mu - 5w_3 - w_4)$$

**Comment:** Note that $\mathrm{BL}(u) = [v_1, v_3, v_2]$ or $\mathrm{BL}(u) = [v_2, v_3, v_1]$ and that $v_1$ and $v_2$ have distinct external neighbors of degree 3.

| 16 | $d(u) = 3, d(v_1) = 3, d(v_2) = 3, d(v_3) = 4, v_1$ and $v_2$ are not adjacent, $v_3$ is adjacent to $v_2$ (or $v_1$) |
|---|---|

$$T(\mu) \geq T(\mu + 2w_2 - 4w_3 - 2w_4) + T(\mu + 2w_2 - 5w_3 - w_4) + T(\mu + 2w_2 - 6w_3 - w_4) + T(\mu + 2w_2 - 7w_3 - w_4)$$

**Comment:** $\mathrm{BL}(u) = [v_2, v_3, v_1]$. The external neighbor of $v_2$ has degree 3 and neighbors of degree 3 and 3 or 4. In the third branch where $v_3$ is selected, $N[v_3]$ is deleted $(-4w_3 - w_4)$, $v_1$ has its degree decreased $(+w_2 - w_3)$, and another vertex has its degree decreased from 3 to 2 $(+w_2 - w_3)$: the external neighbor $x$ of $v_2$ if it is not adjacent to $v_3$, or a neighbor of $x$ if $x$ and $v_3$ are neighbors and $N[x] \not\subseteq N[v_3]$, or the vertex in $N^2(x) \setminus N^2[u]$ in the remaining case.

| 17 | $d(u) = 3, d(v_1) = 3, d(v_2) = 3, d(v_3) = 4, N(u)$ is independent |
|---|---|

$$T(\mu) \geq T(\mu + 2w_2 - 3w_3 - 3w_4) + T(\mu + w_2 - 2w_3 - 3w_4) + T(\mu + w_2 - 2w_3 - 3w_4 - M_2) + T(\mu + 2w_2 - 6w_3 - w_4 - M_2)$$

**Comment:** The external neighbors of $v_1$ and $v_2$ have degree 3 and 3 or 4. In the last two branches, a marked vertex of degree 2 is created.

| 18 | $d(u) = 3, d(v_1) = 3, d(v_2) = 4, d(v_3) = 4, v_1$ is not adjacent to $v_2$ and $v_3$ |
|---|---|

$$T(\mu) \geq T(\mu - w_3 - 3w_4) + T(\mu - 2w_3 - 2w_4) + 2T(\mu + w_2 - 4w_3 - 2w_4)$$

| 19 | $d(u) = 3, d(v_1) = 3, d(v_2) = 4, d(v_3) = 4, v_1$ is adjacent to $v_2$ (or $v_3$) |

$$T(\mu) \geq T(\mu - w_3 - 3w_4) + T(\mu - 2w_3 - 2w_4) + T(\mu - 3w_3 - 2w_4) + T(\mu - 5w_3 - 2w_4)$$

| 20 | $d(u) = 3, d(v_1) = 4, d(v_2) = 4, d(v_3) = 4$ |

$$T(\mu) \geq T(\mu - w_3 - 3w_4) + 3T(\mu - 2w_3 - 3w_4)$$

**Comment:** Consider the branch where $v_1$ is selected. A total of 5 vertices disappear and at least 3 vertices of degree 4 either disappear or have their degree reduced from 4 to 3: the vertices in $N(u)$.

| 21 | $d(u) = 4, d(v_1) = 4, d(v_2) = 4, d(v_3) = 4, d(v_4) = 5$ |

$$T(\mu) \geq 4T(\mu - 5w_4) + T(\mu + 3w_3 - 9w_4)$$

**Comment:** Consider the branch where $v_4$ is selected. A total of 6 vertices disappear and at least 3 vertices have their degree reduced from 4 to 3. We use the same argument for $v_1, v_2$ and $v_3$. Consider $v_1$.
If $v_4$ is not adjacent to $v_1$: the degree of $v_4$ is reduced.
If $v_4$ is adjacent to $v_1$ and $N[v_1] \not\subseteq N[v_4]$: a neighbor of $v_1$ has its degree reduced from 4 to 3.
If $v_4$ is adjacent to $v_1$ and $N[v_1] \subseteq N[v_4]$: Let $y_1$ and $y_2$ be the two common neighbors of $v_1$ and $v_4$ (except $u$). $y_1$ and $y_2$ have degree 4 and neighbors of degree $4, 4, 4$ and 5. At least one of $y_1$ and $y_2$ has a neighbor of degree 4 outside $N[v_4]$, otherwise $N[y_1] = N[y_2]$.

| 22 | $d(u) = 4, d(v_3) = 5, d(v_4) = 5$ |

$$T(\mu) \geq 3T(\mu - 5w_4) + 2T(\mu - 6w_4)$$

| 23 | $d(u) = 4, d(v_4) \geq 6$ |
|---|---|

$$T(\mu) \geq 4T(\mu - 5w_4) + T(\mu - 7w_4)$$

| 24 | $d(u) \geq 5$ |
|---|---|

$$T(\mu) \geq 6T(\mu - 6w_4)$$

### 4.4.5   Count all Maximal Independent Sets in a Marked Graph of Maximum Degree Two

Given a marked graph of maximum degree 2, `#MaximalIS` takes exponential time. We show in this subsection, that all maximal independent sets of a marked graph of maximum degree 2 can be counted in polynomial time. Adding this polynomial time procedure to `#MaximalIS` is likely to be of help in implementations of the algorithm; it does however not improve our analysis of its worst case running time.

Suppose first that $G$ is a path $P_n = (v_1, v_2, \ldots, v_n)$. Let $V_i = \{v_1, v_2, \ldots, v_i\}$ for $i = 1, \ldots, n$. We define three values for the vertices of $G$ with the following meaning:

- $is(v_i)$ - the number of maximal independent sets of $G[V_i]$ containing $v_i$

- $od(v_i)$ - the number of maximal independent sets of $G[V_{i-1}]$ containing $v_{i-1}$

- $ond(v_i)$ - the number of maximal independent sets of $G[V_{i-1}]$ not containing $v_{i-1}$

The algorithm gives the following values to $v_1$:

- $is(v_1) = 0$ if $v_1$ is marked, and 1 otherwise,

- $od(v_1) = 0$, and

- $ond(v_1) = 1$.

Suppose the values for $v_{i-1}$ are known, then the values for $v_i$ are computed by simple dynamic programming as follows:

- $is(v_i) = 0$ if $v_i$ is marked, and $od(v_{i-1}) + ond(v_{i-1})$ otherwise,

- $od(v_i) = is(v_{i-1})$, and

- $ond(v_i) = od(v_{i-1})$.

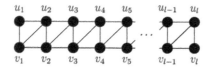

Figure 4.2: Graph $G_l$ used to lower bound the running time of Algorithm #MaximalIS

The number of maximal independent sets satisfying property $\Pi$ (defined in Subsection 4.4.1) of $G$ is $is(v_n) + od(v_n)$.

If $G$ is a cycle $C_n$, select an arbitrary vertex $v_i$ with neighbors $v_{i-1}$ and $v_{i+1}$ and return the sum of the number of maximal independent sets

- containing $v_i$ if $v_i$ is not marked, or 0 otherwise,

- containing $v_{i-1}$ if $v_{i-1}$ is not marked, or 0 otherwise, and

- containing $v_{i+1}$ but not $v_{i-1}$ if $v_{i+1}$ is not marked, or 0 otherwise.

This can easily be done by 3 recursive calls on the instances $G \setminus N[v_i], G \setminus N[v_{i-1}]$ and $G \setminus N[v_{i+1}]$ and by marking $v_{i-1}$ in the last recursive call.

**Lemma 4.7.** *Let $G$ be a marked graph with maximum degree 2. The number of maximal independent sets of $G$ satisfying property $\Pi$ can be computed in linear time.*

*Remark 5.* As $od(v_i) = is(v_{i-1})$, the value $od(\cdot)$ is redundant. But the above description makes it easier to see that a slight generalization of this algorithm, which is very similar to the algorithm in [Alb02], makes it possible to count all maximal independent sets of a marked graph satisfying property $\Pi$ in time $\mathcal{O}(3^k n)$ when a path decomposition of width $k$ of the graph is known.

### 4.4.6 Lower Bound on the Running Time of the Algorithm

For most non-trivial branching algorithms, it is not known whether the upper bound of the running time provided by the currently available analyses is tight or not. A lower bound for the worst case running time of such algorithms is therefore desirable. Here we lower bound the running time of Algorithm #MaximalIS by $\Omega(1.3247^n)$.

**Theorem 4.8.** *There exists an infinite family of graphs for which Algorithm #MaximalIS takes time $\Omega(1.3247^n)$, and thus its worst case running time is $\Omega(1.3247^n)$.*

*Proof.* The lower bound for the running time of #MaximalIS established here uses the same family of graphs as the lower bound for an algorithm computing a minimum independent dominating set [GL06].

Consider the graph $G_l$ of Figure 4.2. It has $n = 2l$ vertices. Note that none of the simplification or halting rules are applicable to $G_l$. The first branching of #MaximalIS is on vertex $u_1$ or vertex $v_l$. Without loss of generality, suppose the algorithm always chooses the vertex with

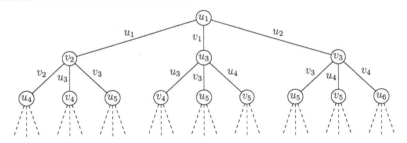

Figure 4.3: A part of the search tree of the execution of Algorithm `#MaximalIS` on the graph $G_l$

smallest index when it has more than one choice (that is it chooses $u_1$ for the first recursive call).

The branching rule (B) then makes recursive calls on graphs with $n-3$, $n-4$ and $n-5$ vertices, not marking any vertex. The structure of all resulting graphs is similar to $G_l$: either isomorphic to $G_{l-2}$ or equal to $G_l \setminus N[u_1]$ or $G_l \setminus N[u_2]$. The subsequent recursive calls again remove 3, 4 and 5 vertices in each case and do not mark any vertices.

The first levels of the corresponding search tree are depicted in Figure 4.3. Unless the graph has at most 4 vertices, each application of branching rule (B) satisfies the recurrence

$$T(n) = T(n-3) + T(n-4) + T(n-5)$$

for this graph and therefore the running time for this class of graphs is $\Omega(\alpha^n)$ where $\alpha$ is the positive root of $x^{-3} + x^{-4} + x^{-5} - 1$, that is $1.3247 < \alpha < 1.3248$. ☐

### 4.4.7 Algorithm to Count all Maximal Bicliques

Finally, we consider the problem of counting all maximal bicliques of a graph $G = (V, E)$. Let $G' = (V', E')$ be a copy of $G$. Let $G'' = (V'', E'')$ where $V'' := V \cup V'$ and $E'' = E \cup E' \cup \{xy' : x, y \in V, y'$ is a copy of $y$ in $V'$, and $(x = y$ or $xy \notin E)\}$.

**Lemma 4.9.** *The number of maximal independent sets of $G''$ equals twice the number of maximal bicliques of $G$.*

*Proof.* We show that there is a one-to-one correspondence between the bicliques of $G$ and the symmetric pairs of independent sets of $G''$.

Let $X \cup Y$ be a biclique of $G$. Clearly, $X, Y$ are independent sets in $G$ and their copies $X', Y'$ are independent sets in $G'$. Let $x \in X$ and $y \in Y$. Then $xy, x'y' \in E''$ and $xy', x'y \notin E''$. So, $X \cup Y'$ and $X' \cup Y$ are independent sets in $G''$.

Let $X, Y \subseteq V$ be such that $X \cup Y'$ is an independent set in $G''$ where $X', Y'$ are the copies of $X, Y$. Hence $X, Y$ are independent sets in $G$. Let $x \in X$ and $y' \in Y'$. Then $xy \in E$. So, $X \cup Y$ is a biclique in $G$. By the symmetry of $G''$, the independent set $X' \cup Y$ in $G''$ also corresponds to the biclique $X \cup Y$ in $G$.

Clearly, this correspondence also holds for maximality by inclusion of vertices.

This implies that $X \cup Y$ is a maximal biclique of $G$ if and only if $X \cup Y'$, and thus also $Y \cup X'$, are maximal independent sets of $G''$.  □

With this construction and the algorithm for counting all maximal independent sets of a graph, we are now able to give an algorithm for counting all maximal bicliques of a graph.

**Theorem 4.10.** *There is an algorithm that counts all maximal bicliques of a given graph $G$ in time $\mathcal{O}(1.3642^n)$, where $n$ is the number of vertices of $G$.*

*Proof.* The algorithm simply calls `#MaximalIS`$((V'', \emptyset, E''))$ and divides the result by 2. Note that $G''$ has $2n$ vertices and that every vertex of $G''$ has degree $n$. The first application of branching rule (B) makes $n+1$ recursive calls and in each one, $n+1$ vertices are removed from the marked graph. Thus the running time is $(n+1)(c^{n-1})n^{\mathcal{O}(1)}$ where $\mathcal{O}^*(c^n)$ is the running time of `#MaximalIS` on a graph with $n$ vertices. The constant $c = 1.3642$ was rounded to derive the running time for `#MaximalIS`, and thus the running time of the algorithm to count maximal bicliques is $\mathcal{O}(1.3642^n)$.  □

## 4.5 Conclusion

We have seen in this chapter that various results for independent sets translate to results for bicliques. But the reverse questions are also interesting. For example, given an algorithm to find a maximum biclique in a graph in time $\mathcal{O}(c^n)$, is it possible to design an $\mathcal{O}(c^n)$ algorithm for MAXIMUM INDEPENDENT SET?

Given a graph $G = (V, E)$ on $n$ vertices, finding a maximum independent set in $G$ could be done by constructing a graph $G'$ obtained from $G$ by adding an independent set $I$ of size $n$ such that every vertex of $I$ is adjacent to every vertex of $V$. Then $G$ has an independent set of size $k$ if and only if $G'$ has a biclique of size $n+k$. This shows that it is possible to obtain a $\mathcal{O}^*(c^{2n})$ algorithm for MAXIMUM INDEPENDENT SET from an algorithm for computing a maximum biclique in a graph in time $\mathcal{O}^*(c^n)$.

A simple variant of this reduction also shows that it is $\mathcal{W}[1]$–hard to find an induced $K_{k,k}$ in a graph, where the parameter is $k$ (now only $k$ independent vertices need to be added to $G$ and made adjacent to every vertex in $V$). However the following question [DGMS07] is still open.

**Open Question.** *Determine the parameterized complexity of the following problem: given a graph $G$ and a parameter $k$, does $G$ have a $K_{k,k}$ as a subgraph.*

Note that the problem in the open question does not require the $K_{k,k}$ to be an *induced* subgraph.

# Chapter 5

# Max 2-Sat, Max 2-CSP, and everything in between

All we ask for is satisfaction.

*Silent Hill (2006), movie*

In this chapter we consider "hybrid" MAX 2-CSP formulae consisting of "simple" clauses, namely conjunctions and disjunctions of pairs of variables, and general 2-variable clauses, which can be any integer–valued functions of pairs of boolean variables. This allows an algorithm to use both efficient reductions specific to AND and OR clauses, and other powerful reductions that require the general CSP setting.

Parametrizing an instance by the fraction $p$ of non-simple clauses, we give an algorithm that is the fastest polynomial space algorithm currently known for MAX 2-SAT (and other $p = 0$ formulae, with arbitrary mixtures of AND and OR clauses); the only efficient algorithm for mixtures of AND, OR, and general integer–valued clauses; and tied for fastest for MAX 2-CSP ($p = 1$). Since a pure 2-SAT input instance may be transformed to a general CSP instance in the course of being solved, the algorithm's efficiency and generality go hand in hand.

The novel analysis in this chapter results in a *family* of running time bounds, each optimized for a particular value of $p$. The algorithm uses new reductions, as well as recent reductions such as "clause–learning" and "2-reductions" adapted to our setting's mixture of simple and general clauses. Each reduction imposes constraints on various parameters, and the running time bound is an "objective function" of these parameters and $p$.

## 5.1 Introduction

**Treatment of "Hybrid" Sat–CSP Formulae.** We show a polynomial space algorithm that solves general instances of integer–valued MAX 2-CSP (formally defined in Section 5.2), but that takes advantage of "simple" clauses, namely unit–weighted conjunctions and disjunctions. In a sense made precise near Remark 7, exclusive–or is the only boolean function we cannot treat efficiently.

| Running Time | Problem | Space | Reference |
|---|---|---|---|
| $\mathcal{O}^*\left(2^{m/2.879}\right)$ | MAX 2-SAT | poly | Niedermeier and Rossmanith [NR00] |
| $\mathcal{O}^*\left(2^{m/3.448}\right)$ | MAX 2-SAT | poly | implicit by Bansal and Raman [BR99] |
| $\mathcal{O}^*\left(2^{m/4}\right)$ | MAX 2-SAT | poly | Hirsch [Hir00] |
| $\mathcal{O}^*\left(2^{m/5}\right)$ | MAX 2-SAT | poly | Gramm et al. [GHNR03] |
| $\mathcal{O}^*\left(2^{m/5}\right)$ | MAX 2-CSP | poly | Scott and Sorkin [SS03] |
| $\mathcal{O}^*\left(2^{m/5.263}\right)$ | MAX 2-CSP | poly | Scott and Sorkin [SS04] |
| $\mathcal{O}^*\left(2^{m/5.217}\right)$ | MAX 2-SAT | poly | Kneis and Rossmanith [KR05] |
| $\mathcal{O}^*\left(2^{m/5.769}\right)$ | MAX 2-SAT | exp | Kneis et al. [KMRR09] |
| $\mathcal{O}^*\left(2^{m/5.5}\right)$ | MAX 2-SAT | poly | Kojevnikov and Kulikov [KK06] |
| $\mathcal{O}^*\left(2^{m/5.769}\right)$ | MAX 2-CSP | exp | Scott and Sorkin [SS07] |
| $\mathcal{O}^*\left(2^{m/5.88}\right)$ | MAX 2-SAT | poly | Kulikov and Kutzkov [KK07] |
| $\mathcal{O}^*\left(2^{m/6.215}\right)$ | MAX 2-SAT | poly | Raible and Fernau [RF08] |
| $\mathcal{O}^*\left(2^{m/5.263}\right)$ | Max 2-CSP | poly | Gaspers and Sorkin [GS09] |
| $\mathcal{O}^*\left(2^{m/6.321}\right)$ | Max 2-Sat | poly | Gaspers and Sorkin [GS09] |

Table 5.1: A historical overview of algorithms for MAX 2-SAT and MAX 2-CSP

Let us give a simple example. In the MAX 2-CSP instance

$$(x_1 \vee x_2) + (x_2 \vee \overline{x_4}) + (x_2 \wedge x_3) + 3 \cdot (x_1 \vee x_3) + (2 \cdot (\overline{x_2}) - 5 \cdot x_4 + (x_2 \oplus x_4)), \qquad (5.1)$$

the first two clauses are unit–weighted disjunctive clauses, the third clause is a unit–weighted conjunction, the fourth clause is a disjunction with weight 3, and the last clause is a general integer–valued CSP clause (any integer–valued 2-by-2 truth table). Thus this example has 3 (the first three clauses) simple clauses and 2 non–simple clauses.

Both MAX 2-SAT and MAX 2-CSP have been extensively studied from the algorithmic point of view. For variable–exponential running times, the only two known algorithms faster than $2^n$ for MAX 2-CSP (and even MAX 2-SAT) are those by Williams [Wil05] and Koivisto [Koi06], both with running time $\mathcal{O}^*\left(2^{n/1.262}\right)$. They employ beautiful ideas, but have exponential space complexity.

For clause–exponential running times, there has been a long series of improved algorithms; see Table 5.1. To solve MAX 2-SAT, all early algorithms treated pure 2-SAT formulae. By using more powerful reductions closed over MAX 2-CSP but not MAX 2-SAT, the MAX 2-CSP generalization of Scott and Sorkin [SS04] led to a faster algorithm. Then, several new MAX 2-SAT specific reductions once again gave the edge to algorithms addressing MAX 2-SAT particularly. Here we get the best of both worlds by using reductions specific to MAX 2-SAT (actually, we allow disjunctive as well as conjunctive clauses), but also using CSP reductions. While it is likely that MAX 2-SAT algorithms will become still faster, we believe that further improvements will continue to use this method of combination.

**Results.** Let $p$ be the fraction of non–simple clauses in the initial instance, no matter how this fraction changes during the execution of the algorithm. In example (5.1), the fraction of non–simple clauses is $p = 2/5$. The algorithm we present here is the fastest known polynomial space algorithm for $p = 0$ (including MAX 2-SAT but also instances with arbitrary mixtures of AND and OR clauses); fastest for all $0 < p < 0.29$ (where no other algorithm is known, short of solving the instance as a case of general MAX 2-CSP); and tied for fastest for $0.29 \leq p \leq 1$, notably for MAX 2-CSP itself. For the well known classes MAX 2-SAT and MAX 2-CSP, our algorithm has running time $\mathcal{O}^* \left( 2^{m/6.321} \right)$ and $\mathcal{O}^* \left( 2^{m/5.263} \right)$, respectively.

For "cubic" instances, where each variable appears in at most three 2-variable clauses, our analysis gives running time bounds that match and generalize the best known when $p = 0$ (including MAX 2-SAT); improve on the best known when $0 < p < 1/2$; and match the best known for $1/2 \leq p \leq 1$ (including MAX 2-CSP).

We derive running time bounds that are optimized to the fraction $p$ of non–simple clauses; see Table 5.2. Every such bound is valid for every formula, but the bound derived for one value of $p$ may not be the best possible for a formula with a different value.

**Method of analysis, and hybrid Sat–CSP formulae.** We view a MAX 2-CSP instance as a constraint graph $G = (V, E \cup H)$ where vertices represent variables, the set of "light" edges $E$ represents simple clauses and the set of "heavy" edges $H$ represents general clauses. The running time analysis of our branching algorithm is measure–based, as described in Chapter 2. The measure $\mu$, that we use to upper bound the running time of the algorithm, includes weights $w_e$ and $w_h$ for each simple and general clause, and weights $w_d$ for each vertex of degree $d$.

To get the best possible running time bound subject to the constraints imposed by our analysis of the reductions, we wish to minimize $\mu(G)$. To avoid looking at the full degree spectrum of $G$, we constrain each vertex weight $w_d$ to be non–positive, and then ignore these terms, resulting in a (possibly pessimistic) running time bound $\mathcal{O}^* \left( 2^{|E|w_e + |H|w_h} \right)$.

If the instance $G$ is a MAX 2-SAT instance, with no heavy edges, to minimize the running time bound is simply to minimize $w_e$ subject to the constraints: as there are no heavy edges in the input instance, it makes no difference if $w_h$ is large. This optimization will yield a small value of $w_e$ and a large $w_h$. Symmetrically, if we are treating a general MAX 2-CSP instance, where all edges are heavy, we need only minimize $w_h$. This optimization will yield weights $w_e, w_h$ that are larger than the MAX 2-SAT value of $w_e$ but smaller than its $w_h$. For a hybrid instance with some edges of each type, minimizing $|E|w_e + |H|w_h$ is equivalent to minimizing $(1 - p)w_e + p w_h$, where $p = |H|/(|E| + |H|)$ is the fraction of non–simple clauses. This will result in weights $w_e$ and $w_h$ each lying between the extremes given by the pure 2-SAT and pure CSP cases; see Figure 5.10 on page 126.

Thus, a new aspect of our approach is that it results in a family of nonlinear programs, not just one: the nonlinear programs differ in their objective functions, which are tuned to the fraction $p$ of non–simple clauses in an input instance. The optimization done for a particular value of $p$, by construction, gives a running time bound that is the best possible (within our methods) for an input instance with this fraction of non–simple clauses. However, it is worth noting that the constraints are the same in all the nonlinear programs, and thus the weights ($w_e$, $w_h$, and the vertex weights) that are optimal for one nonlinear program are feasible for all the nonlinear programs. This means that our family of nonlinear programs results in a family of running time bounds, each of them valid for every input instance, but with the bound optimized to a given value of $p$ being best for formulae with that fraction of non–simple clauses.

**Novel aspects of the analysis.** Our introduction of the notion of hybrids between MAX 2-SAT and MAX 2-CSP, discussed above, is the main distinguishing feature of the present work. It yields a more general algorithm, applicable to CSP instances not just SAT instances, and gives better performance on MAX 2-SAT by allowing both efficient SAT–specific reductions and powerful reductions that go outside that class. This is surely not the final word on MAX 2-SAT algorithms, but we expect new algorithms to take advantage of this hybrid approach.

A secondary point is that CSP reductions such as combining parallel edges or reducing on small cuts mean that in other cases it can be assumed that a graph has no parallel edges or small cuts. This simplifies the case analysis, counter–balancing the complications of considering two types of edges.

The hybrid view marks the biggest change to our approach, since it means that the objective function depends on the fraction of non–simple clauses, so there is a continuum of nonlinear programs, not just one.

Also, it is common to make some assumptions about the weights, but we try to avoid this, instead only limiting the weights by the constraints necessitated by each reduction. This avoids unnecessary assumptions compromising optimality of the result, which is especially important in the hybrid realm where an assumption might be justified for SAT but not for CSP, or vice–versa. It also makes the analysis more transparent.

As is often the case with exact algorithms, regularity of an instance is important, and in our analysis we treat this with explicit weights penalizing regularity (motivated by a similar accounting for the number of 2-edges in a hypergraph, in [Wah04], and by the "forced moves" in [SS07]; see also Subsection 2.8.2 on page 45). This introduces some extra bookkeeping but results in a more structured, more verifiable analysis.

We introduce several new reductions, including a 2-reduction combining ideas from [KK06] (for the SAT case) and [SS07] (the CSP case), a "super 2-reduction", and a generalization of the "clause-learning" from [KK07].

Another useful tool we introduce is a simple graph–theoretic lemma on "good 1-reductions" (Lemma 5.7) which shows that in various situations a certain number of these helpful simplifications occur. This eliminates some amount of case analysis.

## 5.2   Definitions

We use the value 1 to indicate Boolean "true", and 0 "false". The canonical problem MAX SAT is, given a boolean formula in conjunctive normal form (CNF), to find a boolean assignment to the variables of this formula satisfying a maximum number of clauses. MAX 2-SAT is MAX SAT restricted to instances in which each clause contains at most 2 literals.

We will consider a class more general than MAX 2-SAT, namely integer–valued MAX (2,2)-CSP; we will generally abbreviate this to MAX 2-CSP. An instance $(G, S)$ of MAX 2-CSP is defined by a *constraint graph* (or multigraph) $G = (V, E)$ and a set $S$ of *score* functions. There is a *dyadic* score function $s_e \colon \{0,1\}^2 \to \mathbb{Z}$ for each edge $e \in E$, a monadic score function $s_v \colon \{0,1\} \to \mathbb{Z}$ for each vertex $v \in V$, and (for bookkeeping convenience) a single *niladic* score "function" (really a constant) $s_\emptyset \colon \{0,1\}^0 \to \mathbb{Z}$.

A candidate solution is a function $\varphi : V \to \{0,1\}$ assigning values to the vertices, and its score is

$$s(\varphi) := \sum_{uv \in E} s_{uv}(\varphi(u), \varphi(v)) + \sum_{v \in V} s_v(\varphi(v)) + s_\emptyset.$$

An optimal solution $\varphi$ is one which maximizes $s(\varphi)$.

The algorithm we present here solves any instance of MAX 2-CSP with polynomial space usage, but runs faster for instances having a large proportion of "simple" clauses, namely conjunctions and disjunctions.

A hybrid instance $F = (V, E, H, S)$ is defined by its variables or vertices $V$, normal or *light* edges $E$ representing conjunctive clauses and disjunctive clauses, *heavy* edges $H$ representing arbitrary (integer–valued) clauses, and a set $S$ of monadic functions and dyadic functions. Its light–and–heavy–edged constraint graph is $G = (V, E, H)$, though generally we will just think of the graph $(V, E \cup H)$; no confusion should arise. We will write $V(F)$ and $V(G)$ for the vertex set of an instance $F$ or equivalently that of its constraint graph $G$.

We define the degree $d(u)$ of a vertex $u$ to be the number of edges incident on $u$ where loops are counted twice and the *degree* (or *maximum degree*) of a formula $F$ (or its constraint graph $G$) to be the maximum of its vertex degrees. Without loss of generality we will assume that there is at most one score function for each vertex, though we will allow multiple edges. Then, up to constant factors the space required to specify an instance $F$ with constraint graph $G = (V, E, H)$ is the *instance size*

$$|F| := 1 + |V| + |E| + |H|. \tag{5.2}$$

We use the symbol $\boxdot$ to end the description of a reduction rule or the analysis of a case.

## 5.3 Algorithm and Outline of Analysis

We will show an algorithm (sketched as Algorithm `max2csp` in Figure 5.1) which, on input of a hybrid instance $F$, returns an optimal coloring $\varphi$ of $F$'s vertices in time

$$T(F) = \mathcal{O}^* \left( 2^{w_e|E| + w_h|H|} \right). \tag{5.3}$$

### 5.3.1 Algorithm and General Arguments

The central argument (corresponding to the analysis for line 10 of Algorithm `max2csp`) is to establish (5.3) for *simplified* formulae of maximum degree at most 6. We do this shortly, in Lemma 5.1, with the bulk of the chapter devoted to verifying the lemma's hypotheses. Given Lemma 5.1, we then establish a similar running time bound for instances $F$ of degree at most 6 which are not simplified, that is, instances to which we may apply one or more of the simplifications of Procedure `Simplify` (the analysis referred to by line 5 in Algorithm `max2csp`), and for instances of arbitrary degree (the argument alluded to in line 3 of Algorithm `max2csp`).

---

Algorithm max2csp($F$)
**Input** : A hybrid MAX 2-SAT/ CSP instance $F$.
**Output**: An optimal coloring $\varphi$ of the vertices of $F$.

1 **if** $F$ has a vertex $v$ of degree $\geq 7$ **then**
2      Branch on $\varphi(v) = 0$ and $\varphi(v) = 1$ to obtain $F_1$, $F_2$, recursively solve the instances $F_1$
     and $F_2$ and return the best assignment for $F$.
3      (Analysis: Inductively establish running time, using that both $F_1$ and $F_2$ have at
     least 7 edges fewer than $F$.)
4 Simplify $F$. (See procedure Simplify.)
5 (Analysis: Establish running time bound for general instances, using a bound for
simplified instances.)
6 **if** $F$ is nonempty **then**
7      Apply first applicable branching reduction, obtaining $F_1, \ldots, F_k$.
8      Simplify each of $F_1, \ldots, F_k$.
9      Recursively solve $F_1, \ldots, F_k$ and return the best assignment for $F$.
10      (Analysis: Inductively establish running time bound for simplified instances of
     maximum degree $\leq 6$, using $\sum_{i=1}^{k} 2^{\mu(F_i)} \leq 2^{\mu(F)}$.)

Figure 5.1: Outline of Algorithm max2csp and its analysis

---

Procedure Simplify
**Input** : A hybrid instance $F$
**Output**: A simplified instance

**while** Any of the following simplification rules is applicable **do**
     Apply the first applicable simplification: combine parallel edges; remove loops;
     0-reduction; delete a small component; delete a decomposable edge; half–edge
     reduction; 1-reduction; 1-cut; 2-reduction; 2-cut
**return** the resulting simplified instance

Figure 5.2: Procedure Simplify

---

### 5.3.2 Central Argument

The main argument is to establish (5.3) for formulae of maximum degree at most 6. We will
instead prove that

$$T(F) = \mathcal{O}(|F|^k 2^{\mu(F)}), \tag{5.4}$$

which is stronger if (as we will ensure) for some constant $C$ and every simplified instance $F$ of
degree at most 6, the *measure* $\mu(F)$ satisfies

$$\mu(F) \leq w_e|E| + w_h|H| + C. \tag{5.5}$$

The following Lemma follows directly from Lemma 2.5.

**Lemma 5.1** (Main Lemma). *Suppose there exists an algorithm A and constants $D, c \geq 1$, such* *that on input of any hybrid CSP instance $F$ of maximum degree at most $D$, $A$ either solves $F$ directly in time $\mathcal{O}(1)$, or decomposes $F$ into instances $F_1, \ldots, F_k$ all with maximum degree at most $D$, solves these recursively, and inverts their solutions to solve $F$, using time $\mathcal{O}(|F|^c)$ for the decomposition and inversion (but not the recursive solves). Further suppose that for a given measure $\mu$,*

$$(\forall F) \quad \mu(F) \geq 0, \tag{5.6}$$

*and, for any decomposition done by algorithm $A$,*

$$(\forall i) \quad |F_i| \leq |F| - 1, \text{ and} \tag{5.7}$$
$$2^{\mu(F_1)} + \cdots + 2^{\mu(F_k)} \leq 2^{\mu(F)}. \tag{5.8}$$

*Then $A$ solves any instance $F$ of maximum degree at most $D$ in time $\mathcal{O}(|F|^{c+1})2^{\mu(F)}$.*

We will often work with the equivalent to (5.8), that

$$\sum_{i=1}^{k} 2^{\mu(F_i) - \mu(F)} \leq 1. \tag{5.8'}$$

The main work of the chapter will be to find a set of decompositions and a measure $\mu$ such that the decompositions satisfy inequality (5.7), $\mu$ satisfies inequality (5.6), and (more interestingly) $\mu$ satisfies inequality (5.5) for some small values of $w_e$ and $w_h$, and finally, for every decomposition, $\mu$ satisfies inequality (5.8).

### 5.3.3 Measure

For an instance $F$ of (maximum) degree at most 6, we define a measure $\mu(F)$ as a sum of weights associated with light edges, heavy edges, and vertices of various degrees (at most 6), and constants associated with the maximum degree $d$ of $F$ and whether $F$ is regular (for all the degree criteria treating light and heavy edges alike):

$$\mu(F) := \nu(F) + \delta(F), \text{ with} \tag{5.9}$$
$$\nu(F) := |E|w_e + |H|w_h + \sum_{v \in V} w_{d(v)}, \tag{5.10}$$
$$\delta(F) := \sum_{d=4}^{6} \mathsf{K}_\delta(\Delta(G) \geq d)C_d + \sum_{d=4}^{6} \mathsf{K}_\delta(G \text{ is } d\text{-regular})R_d. \tag{5.11}$$

Recall that $\mathsf{K}_\delta(\cdot)$ denotes the logical Kronecker delta.

To satisfy condition (5.5) it is sufficient that

$$(\forall d) \quad w_d \leq 0; \tag{5.12}$$

this is also necessary for large regular instances. Since we are now only considering instances of degree at most 6, we interpret "$\forall d$" to mean for all $d \in \{0, 1, \ldots, 6\}$.

### 5.3.4 Peripheral Arguments

We first dispense with non simplified instances.

**Lemma 5.2.** *Let* $\mathsf{poly}_1(\cdot)$ *and* $\mathsf{poly}_2(\cdot)$ *be two polynomial functions. Suppose that every simplified* MAX *2-CSP instance* $F$ *of degree at most* $D \leq 6$ *can be solved in time* $\mathsf{poly}_1(|F|)2^{\mu(F)}$. *Suppose also that*

1. *simplifying* $F$ *(or determining that* $F$ *is already simplified) takes time at most* $\mathsf{poly}_2(|F|)$,

2. *any instance* $F'$ *obtained from simplifying* $F$ *satisfies* $|F'| \leq |F| - 1$ *and* $\mu(F') \leq \mu(F) + C'$ *for some positive constant* $C'$, *and*

3. *the simplification can be reversed in time at most* $\mathsf{poly}_2(|F|)$ *to recover an optimal solution to* $F$ *from any optimal solution of* $F'$.

*Then every instance* $F$ *of degree at most* $D$ *can be solved in time* $\mathsf{poly}(|F|)2^{\mu(F)}$, *with* $\mathsf{poly}(\cdot) = \mathsf{poly}_2(\cdot) + 2^{C'}\mathsf{poly}_1(\cdot)$.

*Proof.* Since simplifying reduces the instance size, a solution to the original instance $F$ can be obtained in time

$$
\begin{aligned}
T(F) &\leq \mathsf{poly}_2(|F|) + T(F') \\
&\leq \mathsf{poly}_2(|F|) + \mathsf{poly}_1(|F'|)2^{\mu(F')} \\
&\leq \mathsf{poly}_2(|F|) + \mathsf{poly}_1(|F|)2^{\mu(F)+C'} \\
&\leq \left(\mathsf{poly}_2(|F|) + 2^{C'}\mathsf{poly}_1(|F|)\right)2^{\mu(F)} \\
&= \mathsf{poly}(|F|)2^{\mu(F)}.
\end{aligned}
$$

$\square$

The lemma's hypotheses (1) and (3) will be satisfied by construction. Hypothesis (2) is assured if we constrain that, for each simplification rule taking $F$ to $F'$,

$$\nu(F') \leq \nu(F), \tag{5.13}$$

since by transitivity the same inequality then holds for any sequence of simplifications starting with $F$ and ending with a simplified instance $F'$, and the desired inequality $\mu(F') = \nu(F) + \delta(F) - \delta(F') \leq \nu(F) + C'$ follows by boundedness of $\delta$ and choosing $C'$ sufficiently large.

Finally, we dispense with instances of high degree, the argument alluded to in line 3 of Algorithm max2csp.

**Lemma 5.3.** *Suppose that every* MAX *2-CSP instance* $F$ *of degree at most* 6 *can be solved in time* $\mathcal{O}(|F|^{k_1})2^{w_e|E|+w_h|H|}$, *with* $w_e, w_h \geq 1/7$. *Then for some sufficiently large* $k$, *every instance* $F$ *can be solved in time* $\mathcal{O}(|F|^k)2^{w_e|E|+w_h|H|}$.

*Proof.* As in the proof of Lemma 2.3, without loss of generality we may replace the $\mathcal{O}$ statement in the hypothesis with a simple inequality. If $F$ has any vertex $v$ of degree at least 7, we will set $\varphi(v)$ to 0 and 1 to generate instances $F_0$ and $F_1$ respectively, solve them recursively, and note that the solution to $F$ is that of the better of $F_0$ and $F_1$, extended with the corresponding value for $\varphi(v)$. We may assume that the branching and its reversal together take time at most $|F|^{k_2}$.

Ensure that $k \geq k_1$ is large enough that for all $x \geq 2$, $x^{k_2} \leq x^k - (x-1)^k$, and note that the hypothesis remains true replacing $k_1$ with $k$.

The proof is by induction on $F$. If $F$ has no vertex of degree at least 7 then we are already done. Otherwise reduce $F$ to $F_1$ and $F_2$, each having at least 7 fewer (light and/or heavy) edges than $F$. By induction we may assume the bound for $T(F_1)$ and $T(F_2)$, so

$$\begin{aligned}
T(F) &\leq |F|^{k_2} + 2(|F|-1)^k 2^{w_e|E|+w_h|H|-7\cdot 1/7} \\
&= |F|^{k_2} + (|F|-1)^k 2^{w_e|E|+w_h|H|} \\
&\leq |F|^k 2^{w_e|E|+w_h|H|}.
\end{aligned}$$

The worst case for the last inequality is when $w_e|E| + w_h|H| = 0$ (it is nonnegative), and in that case the inequality follows by the construction of $k$. □

### 5.3.5 Optimizing the Measure

The task of the rest of the chapter is to produce the comprehensive set of reductions hypothesized by Lemma 5.1 (to any formula there should be some reduction we can apply) and a measure $\mu$, satisfying the hypotheses, with $w_e$ as small as possible. (More generally, if there are $m(1-p)$ conjunctions and $mp$ general integer–valued clauses, we wish to minimize $m(1-p)w_e + mpw_h$ or equivalently $(1-p)w_e + pw_h$, but for the discussion here we will just think in terms of minimizing $w_e$.)

For each reduction, the hypothesized constraint (5.7) will be trivially satisfied, and it will be straightforward to write down a constraint ensuring (5.8'). We then solve the nonlinear program of minimizing $w_e$ subject to all the constraints.

Minimizing $w_e$ for a given set of constraints can be done with an off–the–shelf nonlinear solver (see Section 5.8.5), but finding a set of reductions resulting in a small value of $w_e$ remains an art. It consists of trying some set of reductions, seeing which ones' constraints are tight in an optimal solution, and trying to replace these reductions with more favorable ones.

With the constraints established in the next sections, we will obtain our main result.

**Theorem 5.4.** *Let $F$ be an instance of integer–weighted* MAX 2-CSP *in which each variable appears in at most $\Delta$ 2-clauses, and there are $(1 - p(F))m$ conjunctive and disjunctive 2-clauses, and $p(F)m$ other 2-clauses. Then, for any pair of values $w_e, w_h$ in Table 5.2, the above algorithm solves $F$ in time $\mathcal{O}^*\left(2^{m\cdot((1-p(F))w_e+p(F)w_h)}\right)$. When the table's $p = p(F)$, we obtain our best bound, $\mathcal{O}^*\left(2^{m\cdot((1-p)w_e+pw_h)}\right) = \mathcal{O}^*\left(2^{mw}\right)$.*

*Proof.* Corollary of Lemma 5.1, solving the mathematical program given by the various constraints given in the next sections and minimizing $pw_h + (1-p)w_e$. □

| $p$ | | 0 | | | 0.05 | | | 0.1 | |
|---|---|---|---|---|---|---|---|---|---|
| $\Delta$ | $w_e$ | $w_h$ | $w$ | $w_e$ | $w_h$ | $w$ | $w_e$ | $w_h$ | $w$ |
| 3 | 0.10209 | 0.23127 | 0.10209 | 0.10209 | 0.23125 | 0.10855 | 0.10209 | 0.23125 | 0.11501 |
| 4 | 0.14662 | 0.31270 | 0.14662 | 0.14662 | 0.31270 | 0.15493 | 0.15023 | 0.26951 | 0.16216 |
| 5 | 0.15518 | 0.30728 | 0.15518 | 0.15637 | 0.27997 | 0.16255 | 0.15640 | 0.27951 | 0.16871 |
| $\geq 6$ | 0.15819 | 0.31029 | 0.15819 | 0.15912 | 0.28223 | 0.16527 | 0.15912 | 0.28223 | 0.17143 |

| $p$ | | 0.2 | | | 0.3 | | | 1 | |
|---|---|---|---|---|---|---|---|---|---|
| $\Delta$ | $w_e$ | $w_h$ | $w$ | $w_e$ | $w_h$ | $w$ | $w_e$ | $w_h$ | $w$ |
| 3 | 0.10209 | 0.23125 | 0.12793 | 0.10209 | 0.23125 | 0.14084 | 0.16667 | 0.16667 | 0.16667 |
| 4 | 0.15023 | 0.26951 | 0.17409 | 0.15023 | 0.26951 | 0.18601 | 0.18750 | 0.18750 | 0.18750 |
| 5 | 0.15640 | 0.27951 | 0.18102 | 0.19000 | 0.19000 | 0.19000 | 0.19000 | 0.19000 | 0.19000 |
| $\geq 6$ | 0.16520 | 0.25074 | 0.18231 | 0.19000 | 0.19000 | 0.19000 | 0.19000 | 0.19000 | 0.19000 |

Table 5.2: Values of $w_e$, $w_h$ and $w := pw_h + (1 - p)w_e$ according to the fraction $p$ of heavy edges and the maximum degree $\Delta$ of a formula $F$. For any pair $(w_e, w_h)$ in the table, a running time bound of $\mathcal{O}^* \left(2^{m \cdot ((1-p)w_e + pw_h)}\right)$ is valid for every formula, regardless of its fraction $p(F)$ of non–simple clauses, but the pair obtained when the table's $p$ equals $p(F)$ gives the best bound

Which of the constraints are tight strongly depends on $p$ and $\Delta$.

## 5.3.6   The Measure's Form

Let us explain the rather strange form of the measure. Ideally, it would be defined simply as $\nu$, and indeed for the measure we ultimately derive, all of our simplifications and most of our branchings satisfy the key inequality (5.8′) with $\nu$ alone in place of $\mu$. Unfortunately, for regular instances of degrees 4, 5, and 6, satisfying this constraint would require a larger value of $w_e$. Viewing (5.8′) equivalently as

$$\sum_{i=1}^{k} 2^{\mu(F_i) - \mu(F)} \leq 1,$$

adding a cost $R_d$ to the measure of a $d$-regular instance $F$ means that if a $d$-regular instance $F$ is reduced to nonregular instances $F_1$ and $F_2$ of degree $d$, each difference $\mu(F_i) - \mu(F)$ is smaller by $R_d$ than the corresponding difference $\nu(F_i) - \nu(F)$ (see Subsection 2.8.2). We will therefore want

$$(\forall d \in \{4, 5, 6\}) \quad R_d \geq 0. \tag{5.14}$$

Of course, if a nonregular instance $F$ of degree $d$ is reduced to instances $F_i$ of degree $d$ one or more of which is regular, there will be a corresponding penalty: for each $d$-regular $F_i$, $\mu(F_i) - \mu(F)$ is $\nu(F_i) - \nu(F) + R_d$.

Indeed, for each branching reduction we will have to consider several cases. Typically, the "baseline" case will be the reduction of a nonregular instance to two nonregular instances. In this case $\mu$ and $\nu$ are equivalent, and if we know for example that $\nu(F_i) - \nu(F) \leq x_i$, our nonlinear program constrains that $2^{x_1} + 2^{x_2} \leq 1$.

If we reduce starting from a regular instance, the nature of the reductions is such that, generically, we will get less favorable bounds $\nu(F_i) - \nu(F) \leq x_i'$ (the values $x_i'$ will be larger than the $x_i$ were), but we also get a "reward" (a further decrease of $R_d$) for whichever of $F_1$ and $F_2$ are not also regular. If we reduce starting from a nonregular instance but producing one or more regular children, we will consider various possibilities.

The cases where a nonregular instance of degree $d$ produces a regular instance $F_i$ of degree at most $d - 1$, can be dispensed with simply by choosing $C_d$ sufficiently large, to reap whatever additional reward is needed. Our branching rules are generally local and will never increase the measure by more than a constant, so some constant $C_d$ suffices. Also, our reductions never increase the degree of an instance (each $F_i$ has degree at most that of $F$), so $C_d$ will never work against us, and there is no harm in choosing it as large as we like. Thus, we never need to consider the particulars of cases where the instance degree decreases, nor the values $C_d$ (see Subsection 2.8.1).

The remaining cases where a nonregular instance has regular children will be considered on a case–by–case basis for each reduction. Generally, for a child to become regular means that, beyond the constraint graph changes taken into account in the baseline case (with the child nonregular), some additional vertices (those of degree less than $d$) must have been removed from the instance by simplifications. Accounting for these implies a further decrease in measure that compensates for the increase by $R_d$.

## 5.4 Some Initial Constraints

We have already derived one constraint for $\mu$, namely (5.12), and we will now introduce some notation and derive several more constraints.

Let us write $w(v)$ for the weight of a vertex $v$ (so $w(v) = w_{d(v)}$), and similarly $w(e)$ for the weight of an edge ($w_e$ or $w_h$ depending on whether $e$ is light or heavy). Sometimes it will be helpful to think of $\nu(F)$ as

$$\nu(F) = \sum_{v \in V} \left( w(v) + \tfrac{1}{2} \sum_{e:\, v \in e} w(e) \right), \tag{5.15}$$

the sum of the weights of the vertices and their incident *half edges*. For convenience, we define (and thus constrain)

$$a_d := w_d + \tfrac{1}{2} d w_e. \tag{5.16}$$

Thus, $a_d$ is equal to the summand in (5.15) for a vertex all of whose incident edges are light, and smaller otherwise.

We require $\mu(F) \geq 0$ for all instances. Considering regular MAX 2-SAT instances with degree $d$ ($d = 0, \ldots, 6$), this implies that

$$(\forall d) \quad a_d \geq 0. \tag{5.17}$$

(For $d \leq 3$, (5.17) is implied by $\delta(F) = 0$, with (5.15) and (5.16). For $d \geq 4$, positivity of $\nu$ might give positive measure to $K_d$ even if $\delta(K_d)$ were negative, but then a graph consisting of sufficiently many copies of $K_d$ would still have negative measure.) If we also constrain that

$$(\forall d \in \{4, 5, 6\}) \quad C_d, R_d \geq 0, \tag{5.18}$$

then we have assured that $\mu(F) \geq 0$ for all instances. In the end, constraint (5.18) will not be tight and so there is no loss in making the assumption.

Were it the case that $w_h \leq w_e$, then we could simply transform each light edge into a heavy one, reducing the measure, and getting a better time bound for solving an instance of MAX 2-CSP than an instance of MAX 2-SAT or a hybrid instance. Thus if we are to gain any advantage from considering MAX 2-SAT or hybrid instances, it must be that

$$w_e \leq w_h. \tag{5.19}$$

In the end we will find that this constraint is not tight, and so there is no cost to making the assumption.[1]

For intuitive purposes let us leap ahead and mention that we will find that $a_0 = a_1 = a_2 = 0$, (thus $w_0 = 0$, $w_1 = -\frac{1}{2}w_e$, and $w_2 = -w_e$), while $0 < a_3 < \cdots < a_6$. Per (5.19) above, $w_h \geq w_e$. Typically we will find that $w_h \leq 2w_e$, but not always. (Even where this fails to hold, notably for cubic MAX 2-SAT, we can still replace two conjunctions or disjunctions on the same variables with one CSP edge: decreasing the degrees of the incident vertices decreases the measure enough to make up for the increase of $w_h - 2w_e$.) This "intuition" has changed several times as the algorithm and its analysis have evolved, which supports the value of making as few assumptions as possible, instead just writing down constraints implied by the reductions.

## 5.5   Simplification Rules and their Weight Constraints

We use a number of simplification rules (reductions of $F$ to a single simpler instance $F_1$ or $F'$). Some of the simplification rules are standard, the CSP 1-reductions are taken from [SS07], the CSP 2-reductions combine ideas from [SS07] and [KK06], and a "super 2-reduction" is introduced here. For vertices of degree 5 we use a branching reduction taken from [KK07] that we generalize to hybrid instances.

We have already ensured constraint (5.6) by (5.17) and (5.18), so our focus is on ensuring that each reduction satisfies (5.8′). Since each branching is followed by an (unpredictable) sequence of simplifications, to have any hope of satisfying (5.8′) it is essential that each simplification from any $F$ to $F'$ satisfies

$$\nu(F') \leq \nu(F); \tag{5.20}$$

in any case this inequality is required by Lemma 5.2 (it duplicates inequality (5.13)). Constraint

---

[1]For the most part we will only write down constraints that are *necessary*, typically being required for some reduction to satisfy (5.8′), but we make a few exceptions early on.

(5.7) of Lemma 5.1 will be trivially satisfied by all our simplifications and branchings.

Recapitulating, in this section we show that (5.20) is satisfied by all our simplifications. Ensuring (5.8′) will come when we look at the branching rules, and the measure component $\delta$ we are ignoring here.

### 5.5.1   Combine Parallel Edges

Two parallel edges (light or heavy) with endpoints $x$ and $y$ may be collapsed into a single heavy edge. This means that the "transformed" instance $F'$ ($F_1$ in Lemma 5.1, with $k = 1$) is identical to $F$ except that the two score functions $s_{xy}(\varphi(x), \varphi(y))$ and $s'_{xy}(\varphi(x), \varphi(y))$ in $F$ are replaced by their sum $s''_{xy}(\varphi(x), \varphi(y))$ in $F'$. If one of the endpoints, say $x$, of the two parallel edges has degree 2, collapse the parallel edges and immediately apply a 1-reduction (see 5.5.7) on $x$ (of degree 1), which removes $x$ from the constraint graph. To ensure (5.20) we constrain

$$(\forall d \geq 2) \quad -a_2 - a_d + a_{d-2} \leq 0: \tag{5.21}$$

the left hand side is $\nu(F') - \nu(F)$ thought of as the subtraction of a vertex of degree 2 and a vertex of degree $d$ and the addition of a vertex of degree $d - 2$. For the case that $x$ and $y$ have degree $d \geq 3$, we constrain

$$(\forall d \geq 3) \quad -2a_d + 2a_{d-1} - w_e + w_h \leq 0: \tag{5.22}$$

the left hand side is $\nu(F') - \nu(F)$ thought of as replacing two vertices of degree $d$ by two vertices of degree $d-1$ and replacing a light edge by a heavy edge. (Remember that the degree of a vertex is the number of incident edges rather than the number of distinct neighbors.) If $d(x) \neq d(y)$, the resulting constraint is a half–half mixture of a constraint (5.22) with $d = d(x)$ and another with $d = d(y)$, and is thus redundant.

By construction, the score functions of $F'$ and $F$ are identical, so an optimal solution $\varphi'$ for $F'$ *is* an optimal solution $\varphi$ of $F'$ (no transformation is needed).                    ⊡

Applying this reduction whenever possible, we may assume that the instance has no parallel edges.

Note that we cannot hope to combine simple clauses (conjunctions and disjunctions) and still take advantage of their being simple clauses rather than general CSP clauses: $(x \vee y) + (\bar{x} \vee \bar{y}) = 1 + (x \oplus y)$, the additive 1 is irrelevant, and the XOR function is not simple.

### 5.5.2   Remove Loops

If the instance includes any edge $xx \in E \cup H$, the nominally dyadic score function $s_{xx}(\varphi(x), \varphi(x))$ may be replaced by a (or incorporated into an existing) monadic score function $s_x(\varphi(x))$. This imposes the constraints

$$(\forall d \geq 2) \quad -a_d + a_{d-2} \leq 0. \tag{5.23}$$

⊡

As this constraint is stronger than (5.21), we may ignore constraint (5.21) now. With this and the edge–combining reduction, we may at all times assume the constraint graph is simple.

### 5.5.3   Delete a Vertex of Degree 0 (0-reduction)

If $v$ is a vertex of degree 0, reduce the instance $F$ to $F'$ by deleting $v$ and its monadic score function $s_v$, solve $F'$, and obtain an optimal solution of $F$ by augmenting the solution of $F'$ with whichever coloring $\varphi(v)$ of $v$ gives a larger value of $s_v(\varphi(v))$. Constraint (5.7) is satisfied, since $|F'| = |F| - 1$. Constraint (5.20) is satisfied if and only if $-w_0 \leq 0$. On the other hand, for a useful result we need each $w_d \leq 0$ (inequality (5.12)), implying that $w_0 = 0$, and thus

$$a_0 = 0. \tag{5.24}$$

We will henceforth ignore vertices of degree 0 completely.                    ⊡

### 5.5.4   Delete a Small Component

For a constant $C$ (whose value we will fix in the branching (reduction 5.7.1)), if the constraint graph $G$ of $F$ has connected components $G'$ and $G''$ with $1 \leq |V(G'')| < C$, then $F$ may be reduced to $F'$ with constraint graph $G'$. The reduction and its correctness are obvious, noting that $F''$ may be solved in constant time. Since $\nu(F') - \nu(F) \leq -\sum_{v \in V(G)} a_{d(v)}$, it is immediate from (5.17) that (5.20) is satisfied.                    ⊡

### 5.5.5   Delete a Decomposable Edge

If a dyadic score function $s_{xy}(\varphi(x), \varphi(y))$ can be expressed as a sum of monadic scores, $s'_x(\varphi(x)) + s'_y(\varphi(y))$, then delete the edge and add $s'_x$ to the original $s_x$, and $s'_y$ to $s_y$. If $x$ and $y$ have equal degrees, the constraint imposed is that $(\forall d \geq 1) \; -w_e - 2w_d + 2w_{d-1} \leq 0$, or equivalently,

$$(\forall d \geq 1) \quad - a_d + a_{d-1} \leq 0 \tag{5.25}$$

(the $d = 1$ case was already implied by (5.24) and (5.17)). As in (5.22), inequalities for degree pairs are a mixture of those for single degrees. Note that we may ignore constraint (5.23) now as it is weaker than (5.25).                    ⊡

Three remarks. First, together with (5.24), (5.25) means that

$$0 = a_0 \leq a_1 \leq \cdots \leq a_6. \tag{5.26}$$

Second, if an edge is not decomposable, the assignment of either endpoint has a (nonzero) bearing on the optimal assignment of the other, as we make precise in Remark 6. We will exploit this in Lemma 5.5, which shows how "super 2-reduction" opportunities (reduction 5.6.1) are created.

*Remark* 6. Let

$$\text{bias}_y(i) := s_{xy}(i,1) - s_{xy}(i,0),$$

the "preference" of the edge function $s_{xy}$ for setting $\varphi(y) = 1$ over $\varphi(y) = 0$ when $x$ is assigned $\varphi(x) = i$. Then $s_{xy}$ is decomposable if and only if $\text{bias}_y(0) = \text{bias}_y(1)$.

*Proof.* $s_{xy}$ is decomposable if and only if its 2-by-2 table of function values has rank 1, which is equivalent to equality of the two diagonal sums, $s_{xy}(0,1)+s_{xy}(1,0) = s_{xy}(0,0)+s_{xy}(1,1)$, which in turn is equivalent to $s_{xy}(0,1)-s_{xy}(0,0) = s_{xy}(1,1)-s_{xy}(1,0)$, that is, $\text{bias}_y(0) = \text{bias}_y(1)$. □

Finally, when some vertices and their incident edges are deleted from a graph, we may think of this as the deletion of each vertex and its incident half–edges (which typically we will account for explicitly) followed (which we may not account for) by the deletion of any remaining half–edges and the concomitant decrease in the degrees of their incident vertices (for edges one of whose endpoints was deleted and one not). A "half–edge deletion" and vertex degree decrease is precisely what is characterized by the left–hand side of (5.25), so it cannot increase the measure $\nu$. Even though such simplifications take place on an intermediate structure that is more general than a graph, and that we will not formalize, for convenient reference we will call this a half–edge reduction.

## 5.5.6   Half–Edge Reduction

Delete a half–edge, and decrease the degree of its incident vertex. By (5.25), this does not increase the measure.

## 5.5.7   Delete a Vertex of Degree 1 (1-reduction)

This reduction comes from [SS07], and works regardless of the weight of the incident edge. Let $y$ be a vertex of degree 1, with neighbor $x$. Roughly, we use the fact that the optimal assignment of $y$ is some easily computable function of the assignment of $x$, and thus $y$ and its attendant score functions $s_y(\varphi(y))$ and $s_{xy}(\varphi(x), \varphi(y))$ can be incorporated into $s_x(\varphi(x))$.

We take a precise formulation from [SS07]. Here $V$ is the vertex set of $F$, $E$ is the set of all edges (light and heavy), and $S$ is the set of score functions.

Reducing $(V, E, S)$ on $y$ results in a new instance $(V', E', S')$ with $V' = V \backslash y$ and $E' = E \backslash xy$. $S'$ is the restriction of $S$ to $V'$ and $E'$, except that for all "colors" $C \in \{0,1\}$ we set

$$s'_x(C) := s_x(C) + \max_{D \in \{0,1\}} \{s_{xy}(CD) + s_y(D)\}.$$

Note that any coloring $\varphi'$ of $V'$ can be extended to a coloring $\varphi$ of $V$ in two ways, depending on the color assigned to $y$. Writing $(\varphi', D)$ for the extension in which $\varphi(y) = D$, the defining property of the reduction is that $S'(\varphi') = \max_D S(\varphi', D)$. In particular, $\max_{\varphi'} S'(\varphi') = \max_\varphi S(\varphi)$, and an optimal coloring $\varphi'$ for the instance $(V', E', S')$ can be extended to an optimal coloring $\varphi$ for $(V, E, S)$. This establishes the validity of the reduction.

Since the reduction deletes the vertex of degree 1 and its incident edge (light, in the worst case), and decreases the degree of the adjacent vertex, to ensure (5.20), we constrain that $(\forall d \geq 1) -w_1 - w_e - w_d + w_{d-1} \leq 0$, or equivalently that

$$(\forall d \geq 1) \quad a_{d-1} - a_d - a_1 \leq 0,$$

which is already ensured by constraint (5.26).           $\boxdot$

### 5.5.8    1-cut

Let $x$ be a cut vertex isolating a set of vertices $A$, $2 \leq |A| \leq 10$. (The 1-cut reduction extends the 1-reduction, thought of as the case $|A| = 1$.) Informally, for each of $\varphi(x) = 0, 1$ we may determine the optimal assignments of the vertices in $A$ and the corresponding optimal score; adding this score function to the original monadic score $s_x$ gives an equivalent instance $F'$ on variables $V \setminus A$. With $A$ of bounded size, construction of $F'$, and extension of an optimal solution of $F'$ to one of $F$, can be done in polynomial time. (Formal treatment of a more general "cut reduction" on more general "Polynomial CSPs" can be found in [SS09].)

This simplification imposes no new constraint on the weights. Vertices in $A$ and their incident half–edges are deleted, and any remaining half–edges (those incident on $x$) are removed by half–edge reductions (reduction 5.5.6); by (5.26), neither increases the measure $\nu$.    $\boxdot$

### 5.5.9    Contract a Vertex of Degree 2 (2-reduction)

Let $y$ be a vertex of degree 2 with neighbors $x$ and $z$. Then $y$ may be contracted out of the instance: the old edges $xy$, $yz$, and (if any) $xz$ are replaced by a single new edge $xz$ which in general is heavy, but is light if there was no existing edge $xz$ and at least one of $xy$ and $yz$ was light.

The basics are simple, but care is needed both because of the distinction between light and heavy edges and because we insist that the constraint graph be simple, and the 2-reduction is the one operation that has the capacity to (temporarily) create parallel edges and in the process change the vertex degrees. We consider two cases: there is an edge $xz$; and there is no edge $xz$.

If there is an edge $xz$ then $x$ and $z$ both have degree 3 or more by Simplification 5.5.8, we use the general MAX 2-CSP 2-reduction from [SS07]. Arguing as in the 1-reduction above, here the optimal assignment of $y$ depends only on the assignments of $x$ and $z$, and thus we may incorporate all the score terms involving $y$, namely $s_y(\varphi(y)) + s_{xy}(\varphi(x), \varphi(y)) + s_{yz}(\varphi(y), \varphi(z))$, into a new $s'_{xz}(\varphi(x), \varphi(z))$, which is then combined with the original $s_{xz}(\varphi(x), \varphi(z))$. The effect is that $y$ is deleted, three edges (in the worst case all light) are replaced by one heavy edge, and the degrees of $x$ and $z$ decrease by one. If $d(x) = d(y) = d$, $\nu(F') - \nu(F) \leq 0$ is assured by $-w_2 - 3w_e + w_h - 2w_d + 2w_{d-1} \leq 0$, or equivalently

$$(\forall d \geq 3) \quad -a_2 - w_e + w_h - 2a_d + 2a_{d-1} \leq 0,$$

which is already ensured by (5.22) and (5.17). As in (5.25), inequalities for pairs $d(x) \neq d(y)$

are a mixture of those for single degrees. *If $xy$ or $yz$ is heavy, then $\nu(F') - \nu(F) \leq -w_h + w_e$, and we will capitalize on this later.*

Finally, we consider the case where there was no edge $xz$. If $xy$ and $yz$ are both heavy, then as in the first case we apply the general MAX 2-CSP reduction to replace them with a heavy edge $xz$, giving $\nu(F') - \nu(F) \leq -2w_h + w_h - w_2 = -a_2 - w_h + w_e \leq -w_h + w_e$.

Otherwise, at least one of $xy$ and $yz$ is light, and we show that the resulting edge $xz$ is light. (For pure SAT formulae, this is the "frequently meeting variables" rule of [KK06].) Without loss of generality we assume that $xy$ is the conjunctive constraint $x \vee y$ or the disjunction $x \wedge y$ (what is relevant is that the clause's score is restricted to $\{0, 1\}$, and is monotone in $\varphi(y)$). We define a *bias*

$$\text{bias}_y(i) := (s_y(1) - s_y(0)) + (s_{yz}(1, i) - s_{yz}(0, i)), \tag{5.27}$$

to be the "preference" (possibly negative) of $s_y + s_{yz}$ for setting $\varphi(y) = 1$ versus $\varphi(y) = 0$, when $z$ has been assigned $\varphi(z) = i$. If $\text{bias}_y(i) \leq -1$ then $\varphi(y) = 0$ is an optimal assignment. (That is, for every assignment to the remaining variables, including the possibility that $\varphi(x) = 0$, setting $\varphi(y) = 0$ yields at least as large as score as $\varphi(y) = 1$.) Also, if $\text{bias}_y(i) \geq 0$ then $\varphi(y) = 1$ is an optimal assignment.

Thus, an optimal assignment $\varphi(y)$ can be determined as a function of $\varphi(z)$ alone, with no dependence on $\varphi(x)$. (This cannot be done in the general case where $xy$ and $yz$ are both heavy edges.) With $\varphi(y)$ a function of $\varphi(z)$, the score $s_{yz}(\varphi(y), \varphi(z))$ may be incorporated into the monadic score function $s_z(\varphi(z))$. Also, there are only 4 functions from $\{0, 1\}$ to $\{0, 1\}$: as a function of $\varphi(z)$, $\varphi(y)$ must the constant function 0 or 1 (in which cases $x \vee y$ can be replaced respectively by a monadic or niladic clause) or $\varphi(z)$ or $\overline{\varphi(z)}$ (in which cases $x \vee y$ can be replaced respectively by the Sat clause $x \vee z$ or $x \vee \bar{z}$).

This shows that if there is no edge $xz$ and either $xy$ or $yz$ is light, then the 2-reduction produces a light edge $xz$. If both $xy$ and $yz$ are light, $\nu(F') - \nu(F) \leq -a_2 \leq 0$, while (once again) if one of $xy$ and $yz$ is heavy, $\nu(F') - \nu(F) \leq -w_h + w_e$.

To summarize, no new constraint is imposed by 2-reductions. Also, if either of $xy$ or $yz$ is heavy, then we have not merely that $\nu(F') - \nu(F) \leq 0$ but that $\nu(F') - \nu(F) \leq -w_h + w_e$, and we will take advantage of this later on.                    ⊡

### 5.5.10    2-cut

Let $\{x, y\}$ be a 2-cut isolating a set of vertices $A$, $2 \leq |A| \leq 10$. (The 2-cut reduction extends the 2-reduction, thought of as the case $|A| = 1$.) Similarly to the 1-cut above, for each of the four cases $\varphi : \{x, y\} \rightarrow 0, 1$ we may determine the optimal assignments of the vertices in $A$ and the corresponding optimal score; adding this score function to the original dyadic score $s_{xy}$ gives an equivalent instance $F'$ on variables $V \setminus A$. There is nothing new in the technicalities, and we omit them.

In general, $\nu' - \nu$ may be equated with the weight change from deleting the original edge $xy$ if any (guaranteed by (5.25) not to increase the measure), deleting all vertices in $A$ with their incident half edges (a change of $-\sum_{v \in A} a_{d(v)}$), replacing one half–edge from each of $x$

and $y$ into $A$ with a single heavy edge between $x$ and $y$ (not affecting their degrees, and thus a change of $-w_e + w_h$), then doing half–edge reductions to remove any half–edges remaining from other edges in $\{x, y\} \times A$ (guaranteed by reduction 5.5.6 not to increase the measure). Thus, $-\sum_{v \in A} a_{d(v)} - w_e + w_h \le -2a_3 - w_e + w_h$, where the second inequality uses that $|A| \ge 2$, all vertices have degree $\ge 3$ (a 2-reduction is preferred to this 2-cut reduction), and the values $a_i$ are nondecreasing (see (5.26)). Thus we can assure that $\nu' - \nu \le 0$ by

$$-2a_3 - w_e + w_h \le 0,$$

which is already imposed by (5.22) and (5.17).                                                    ⊡

## 5.6  Some Useful Tools

Before getting down to business, we remark that in treating disjunction and conjunction efficiently, as well as decomposable functions (see reduction 5.5.5 and Remark 6), the only boolean function our algorithm cannot treat efficiently is exclusive–or. The following remark is surely well known.

*Remark 7.* The only non decomposable two-variable boolean functions are conjunction, disjunction, and exclusive–or.

*Proof.* A function $s\colon \{0,1\}^2 \mapsto \{0,1\}$ is characterized by a 2-by-2 table of 0s and 1s. If the table has rank 1 (or 0), we can decompose $s$ into monadic functions writing $s_{xy}(\varphi(x), \varphi(y)) = s_x(\varphi(x)) + s_y(\varphi(y))$. A table with zero or four 1s is a constant function, trivially decomposable. A table with one 1 is the function $\varphi(x) \wedge \varphi(y)$, up to symmetries of the table and (correspondingly) negations of one or both variables; similarly a table with three 1s is the function $\varphi(x) \vee \varphi(y)$. In a table with two 1s, either the 1s share a row or column, in which case the function is decomposable, or they lie on a diagonal, which is (up to symmetries and signs) the function $\varphi(x) \oplus \varphi(y)$.                                                    □

The property of disjunction and conjunction on which we rely (besides having range $\{0,1\}$) is that they are monotone in each variable. Obviously exclusive–or is not monotone, and it seems that it cannot be accommodated by our methods.

### 5.6.1  Super 2-reduction

Suppose that $y$ is of degree 2 and that its optimal color $C \in \{0,1\}$ is independent of the colorings of its neighbors $x$ and $z$, that is,

$$(\forall D, E) \quad s_y(C) + s_{yx}(C, D) + s_{yz}(C, E) = \max_{C' \in \{0,1\}} s_y(C') + s_{yx}(C', D) + s_{yz}(C', E). \quad (5.28)$$

In that case, $s_y(\varphi(y))$ can be replaced by $s_y(C)$ and incorporated into the niladic score, $s_{xy}(\varphi(x), \varphi(y))$ can be replaced by a monadic score $s'_x(\varphi(x)) := s_{xy}(\varphi(x), C)$ and combined with the existing $s_x$, and the same holds for $s_{yz}$, resulting in an instance with $y$ and its incident edges deleted.                                                                                       ⊡

A super 2-reduction is better than a usual one since $y$ is deleted, not just contracted.

We will commonly branch on a vertex $u$, setting $\varphi(u) = 0$ and $\varphi(u) = 1$ to obtain instances $F_0$ and $F_1$, and solving both.

**Lemma 5.5.** *After branching in a simplified instance $F$ on a vertex $u$ incident to a vertex $y$ of degree 3 whose other two incident edges $xy$ and $yz$ are both light, in at least one of the reduced instances $F_0$ or $F_1$, $y$ is subject to a super 2-reduction.*

*Proof.* In the clauses represented by the light edges $xy$ and $yz$, let $b \in \{-2, 0, 2\}$ be the number of occurrences of $y$ minus the number of occurrences of $\bar{y}$. (As in reduction 5.5.9, we capitalize on the fact that conjunction and disjunction are both elementwise monotone, and that their scores are limited to $\{0, 1\}$.) Following the fixing of $u$ to 0 or 1 and its elimination, let $\text{bias}_y := s_y(1) - s_y(0)$. Given that $F$ was simplified, the edge $uy$ was not decomposable, so by Remark 6 the value of $\text{bias}_y$ in $F_0$ is unequal to its value in $F_1$.

First consider the case $b = 0$. If $\text{bias}_y \geq 1$, the advantage from $\text{bias}_y$ for setting $\varphi(y) = 1$ rather than 0 is at least equal to the potential loss (at most 1) from the one negative occurrence of $y$ in $xy$ and $yz$, so the assignment $\varphi(y) = 1$ is always optimal. Symmetrically, if $\text{bias}_y \leq -1$ we may set $\varphi(y) = 0$. The only case where we cannot assign $y$ is when $\text{bias}_y = 0 = -b/2$.

Next consider $b = 2$. (The case $b = -2$ is symmetric.) If $\text{bias}_y \geq 0$ we can fix $\varphi(y) = 1$, while if $\text{bias}_y \leq -2$ we can fix $\varphi(y) = 0$. The only case where we cannot assign $y$ is when $\text{bias}_y = -1 = -b/2$.

Thus, we may optimally assign $y$ independent of the assignments of $x$ and $z$ unless $\text{bias}_y = -b/2$. Since $\text{bias}_y$ has different values in $F_0$ and $F_1$, in at least one case $\text{bias}_y \neq -b/2$ and we may super 2-reduce on $y$.                                                                      □

## 5.6.2    Branching on Vertices of Degree 5

Kulikov and Kutzkov [KK07] introduced a clever branching on vertices of degree 5. Although we will not use it until we address instances of degree 5 in Section 5.10, we present it here since the basic idea is the same one that went into our 2-reductions: that in some circumstances an optimal assignment of a variable is predetermined. In addition to generalizing from degree 3 to degree 5 (from which the generalization to every degree is obvious), [KK07] also applies the idea somewhat differently.

The presentation in [KK07] is specific to 2-SAT. Reading their result, it seems unbelievable that it also applies to MAX 2-CSP as long as the vertex being reduced upon has only light incident edges (even if its neighbors are incident to heavy edges), but in fact the proof carries over unchanged.

**Lemma 5.6** (clause learning). *In a MAX 2-CSP instance $F$, let $u$ be a variable of degree 5, with light edges only, and neighbors $v_1, \ldots, v_5$. Then there exist "preferred" colors $C_u$ for $u$ and $C_i$ for each neighbor $v_i$ such that a valid branching of $F$ is into three instances: $F_1$ with*

$\varphi(u) = C_u$; $F_2$ with $\varphi(u) \neq C_u$, $\varphi(v_1) = C_1$; and $F_3$ with $\varphi(u) \neq C_u$, $\varphi(v_1) \neq C_1$, and $\varphi(v_i) = C_i$ ($\forall i \in \{2, 3, 4, 5\}$).

*Proof.* For any coloring $\varphi : V \to \{0, 1\}$, let $\varphi^0$ and $\varphi^1$ assign colors 0 and 1 respectively to $u$, but assign the same colors as $\varphi$ to every other vertex. That is, $\varphi^i(u) = i$, and ($\forall x \neq u$) $\varphi^i(x) = \varphi(x)$.

What we will prove is that for any assignment $\varphi$ in which at least two neighbors of $u$ do not receive their preferred colors, $s(\varphi^{C_u}) \geq s(\varphi)$: the assignment in which $u$ receives its preferred color has score at least as large as that in which it receives the other color, and thus we may exclude the latter possibility in our search. (This may exclude some optimal solutions, but it is also sure to retain an optimal solution; thus this trick will not work for counting, but does work for optimization.) That is, if $u$ and one neighbor (specifically, $v_1$) do not receive their preferred color, then we may assume that every other neighbor receives its preferred color.

It suffices to show the existence of colors $C_u$ and $C_i$, $i \in 1, \ldots, 5$, such that for any $\varphi$ with $\varphi(i) \neq C_i$ for two values of $i \in \{1, \ldots, 5\}$, we have $s(\varphi^{C_u}) \geq s(\varphi)$.

Leave the immediate context behind for a moment, and consider any MAX 2-CSP instance $F$ in which some variable $u$ has only light edges, and in them appears $N_2^+$ times positively and $N_2^-$ times negatively. (As in reduction 5.5.9 and Lemma 5.5, we are using the fact that conjunction and disjunction are elementwise monotone.) If $\varphi(u) = 0$, the total score $s^0$ from terms involving $u$ satisfies

$$s_u(0) + N_2^- \leq s^0 \leq s_u(0) + N_2^- + N_2^+,$$

and if $\varphi(u) = 1$ the corresponding score $s^1$ satisfies

$$s_u(1) + N_2^+ \leq s^1 \leq s_u(1) + N_2^+ + N_2^-.$$

From the second inequality in the first line and the first inequality in the second line, if $s_u(1) - s_u(0) \geq N_2^-$ then $s^1 \geq s^0$, and for any coloring $\varphi$, $s(\varphi^1) \geq s(\varphi^0)$. Symmetrically, if $s_u(0) - s_u(1) \geq N_2^+$ then $\varphi^0$ always dominates $\varphi^1$. Defining the bias

$$b := s_u(1) - s_u(0),$$

we may thus infer an optimal color for $u$ if $b - N_2^- \geq 0$ or $-b - N_2^+ \geq 0$.

If $u$ has degree 5, $(b - N_2^-) + (-b - N_2^+) = -N_2^- - N_2^+ = -5$, and thus one of these two parenthesized quantities must be at least $-2.5$, and by integrality at least $-2$. Given the symmetry, without loss of generality suppose that $b - N_2^- \geq -2$. The preferred color for $u$ will be $C_u = 1$.

A small table shows that for any conjunctive or disjunctive clause involving $u$ or $\bar{u}$ and some other variable $v_i$ (which without loss of generality we assume appears positively), there exists a color $C_i$ for $v_i$ (according to the case) such that assigning $v_i$ this color increases $b - N_2^-$ by 1 (either by increasing the bias and leaving $N_2^-$ unchanged, or leaving the bias unchanged and decreasing $N_2^-$).

| original clause | set $\varphi(v_i) =$ $C_i =$ | resulting clause | change in $b$ | change in $N_2^-$ | change in $b - N_2^-$ |
|---|---|---|---|---|---|
| $(u \vee v_i)$ | 0 | $(u)$ | $+1$ | 0 | $+1$ |
| $(u \wedge v_i)$ | 1 | $(u)$ | $+1$ | 0 | $+1$ |
| $(\bar{u} \vee v_i)$ | 1 | $(1)$ | 0 | $-1$ | $+1$ |
| $(\bar{u} \wedge v_i)$ | 0 | $(0)$ | 0 | $-1$ | $+1$ |

Thus, starting from $b - N_2^- \geq -2$, assigning to any two neighbors of $u$ their color $C_i$ results in an instance in which $b - N_2^- \geq 0$, and thus in which an optimal assignment for $u$ is $\varphi(u) = C_u = 1$. This proves the lemma. $\qquad\square$

### 5.6.3 A Lemma on 1-reductions

A half–edge reduction or 1-reduction is "good" if the target vertex has degree at lest 3, because (as the weights will come out) the measure decrease due to $a_{d-1} - a_d$ is substantial for $d \geq 3$, but small (in fact, 0) for $d = 1$ and $d = 2$.

If for example we start with a simplified instance (in which all vertices must have degree at least 3) and reduce on a vertex of degree d, deleting it and its incident half–edges, each of the $d$ remaining half–edges implies a good degree reduction on a neighboring vertex. However, if we deleted several vertices, this might not be the case: if two deleted vertices had a common neighbor of degree 3, its degree would be reduced from 3 to 2 by one half–edge reduction (good), but then from 2 to 1 by the other (not good).

The following lemma allows us to argue that a certain number of good half–edge reductions occur. The lemma played a helpful role in our thinking about the case analysis, but in the presentation here we invoke it rarely: the cases dealt with are relatively simple, and explicit arguments are about as easy as applying the lemma.

Note that for any half edge incident on a vertex $v$, we can substitute a full edge between $v$ and a newly introduced vertex $v'$: after performing a half–edge reduction on $v$ in the first case or a 1-reduction in the second, the same instance results. (Also, the measure increase of $a_1$ when we add the degree-1 vertex and half–edge is canceled by the extra decrease for performing a 1-reduction rather than a half–edge reduction.) For clarity of expression, the lemma is thus stated in terms of graphs and 1-reductions, avoiding the awkward half edges.

**Lemma 5.7.** *Let $G$ be a graph with $k$ degree-1 vertices, $X = \{x_1, \ldots, x_k\}$. It is possible to perform a series of 1-reductions in $G$ where each vertex $x_i$ in $X$ is either matched one-to-one with a good 1-reduction (a 1-reduction on a vertex of degree 3 or more), or belongs to a component of $G$ containing at least one other vertex of $X$, where the total order of all such components is at most $2k$ plus the number of degree-2 vertices.*

In particular, if $G$ is a connected graph then there are $k$ good 1-reductions. By analogy with the well–definedness of the 2-core of a graph, any series of 1-reductions should be equivalent, but the weaker statement in the lemma suffices for our purposes.

*Proof.* The intuition is that each series of reductions originating at some $x_i \in X$, after propagating through a series of vertices of degree 2, terminates either at a vertex of degree 3 or more (reducing its degree), establishing a matching between $x$ and a good reduction, or at another vertex $x_j \in X$, in which case the path from $x_i$ to $x_j$ (or some more complex structure) is a component.

Starting with $i = 1$, let us 1-reduce from $x_i$ as long as possible before moving on to $x_{i+1}$. That is, if we 1-reduce into a vertex of degree 2 we perform a new 1-reduction from that vertex, terminating when we reach a vertex of degree 1 or degree 3 or more. Rather than deleting an edge with a 1-reduction, imagine that the edges are originally black, and each reduced edge is replaced by a red one (which of course is not available for further 1-reductions).

We assert that just before we start processing any $x_i$, the red–edged graph has components consisting of vertices *all* of whose edges are red (in which case this is also a component in $G$ itself), and components where all vertices but one *component owner* are all–red, and the component owner has at least 1 red edge and at least 2 black edges. We prove this by induction on $i$, with $i = 1$ being trivial.

Given that it is true before $x_i$, we claim that: (1) as we reduce starting with $x_i$, the reduction sequence is uniquely determined; (2) in the red–edged component including $x_i$, all vertices are all–red except for a single *active* one; and (3) the sequence on $x_i$ ends when we reduce a vertex that had at least 3 black edges (matching $x_i$ with this good reduction), or a vertex $x_j \in X, j > i$ (in which case we will show that the red component including $x_i$ and $x_j$ is also a component of $G$ itself).

We prove these claims by induction on the step number, the base case again being trivial ($x_i$ itself is active). If we reduce into a vertex $v$ with two black edges (we will say it has *black degree* 2), the next reduction takes us out its other black edge, leaving both red. If $v$ was of degree 2 it is added to $x_i$'s red component; if not, it must have been a component owner (these are the only mixed–color vertices), and we unite the vertex and its component with $x_i$'s component. If we reduce into a vertex $v$ with at least 3 black edges, we match $x_i$ with the good reduction on $v$, and $v_i$ owns $x_i$'s red component. The only remaining possibility is that we reduce into a vertex with 1 black edge, which can only be a degree-1 vertex $x_j$ (with $j > i$), as there are no mixed–color vertices with 1 black edge. In this case we add $x_j$ to $x_i$'s component, and terminate the sequence of reductions for $x_i$ without a good reduction. However the red component on $x_i$ now has no black edges on any of its vertices, and is thus a component in the original black graph $G$.

Starting with the $k$ vertices $x_i$ as initial red components, as we generate the component for $x_i$, the union of all components is expanded as we pass through (and use up) a (true) degree-2 vertex, left unchanged if we pass through a vertex of higher degree with black degree 2, expanded as we enter a terminal all–black degree-3 vertex, and left unchanged if we terminate at another vertex $x_j$. Then, recalling that $k$ is the number of degree-1 vertices in $X$ and letting $k_2$ be the number of degree-2 vertices, the total number of vertices in the union of all components is at most the number of seeds ($k$), plus the number of pass–throughs (at most $k_2$), plus the number of good terminals (at most $k$). In particular, we can partition $X$ into the set of vertices with good terminals in $G$, and the rest; the rest lie in components of $G$ where the total size of these components is $\leq 2k + k_2$.    □

## 5.7  Branching Reductions and Preference Order

Recall from Algorithm `max2csp` that if we have a nonempty simplified instance $F$, we will apply a branching reduction to produce smaller instances $F_1, \ldots, F_k$, simplify each of them, and argue that $\sum_{i=1}^{k} 2^{\mu(F_i)-\mu(F)} \leq 1$ (inequality (5.8′)).

We apply branching reductions in a prescribed order of preference, starting with division into components.

### 5.7.1  Split large components

If the constraint graph $G$ of $F$ has components $G_1$ and $G_2$ with at least $C$ vertices each ($C$ is the same constant as in the simplification rule (5.5.4)), decompose $F$ into the corresponding instances $F_1$ and $F_2$. The decomposition is the obvious one: monadic score functions $s_x$ of $F$ are apportioned to $F_1$ or $F_2$ according to whether $x$ is a vertex of $G_1$ or $G_2$, similarly for dyadic score functions and edges $xy$, while we may apportion the niladic score function of $F$ to $F_1$, setting that of $F_2$ to 0.

It is clear that this is a valid reduction, but we must show that (5.8′) is satisfied. Note that $\nu(F_1) + \nu(F_2) = \nu(F)$, and $\nu(F_i) \geq Ca_3$ since $F_i$ has at least $C$ vertices, all degrees are at least 3, and the $a_i$ are nondecreasing. Thus $\nu(F_1) \leq \nu(F) - Ca_3$. Also, $\delta(F_1) - \delta(F)$ is constant–bounded. Assuming that $a_3 > 0$, then for $C$ sufficiently large,

$$\mu(F_1) - \mu(F) = \nu(F_1) - \nu(F) + \delta(F_1) - \delta(F)$$
$$\leq -Ca_3 + \sum_{d=4}^{6} (R_d + C_d)$$
$$\leq -1.$$

The same is of course true for $F_2$, giving $2^{\mu(F_1)-\mu(F)} + 2^{\mu(F_2)-\mu(F)} \leq 2^{-1} + 2^{-1} = 1$ as required.

The non strict inequality $a_3 \geq 0$ is established by (5.17), and if $a_3 = 0$, a 3-regular (cubic) instance would have measure 0, implying that we could solve it in polynomial time, which we do not know how to do. Thus let us assume for a moment that

$$a_3 > 0. \tag{5.29}$$

This strict inequality (in fact $a_3 \geq 1/7$) will be implied by the constraints for branching rules for cubic instances, constraint (5.31) for example.    ⊡

If $F$'s constraint graph is connected the branching we apply depends on the degree of $F$, that is, the degree of its highest–degree vertex. Although high–degree cases thus take precedence, it is easier to discuss the low–degree cases first. Sections 5.8, 5.9, 5.10, and 5.11 detail the branchings for (respectively) instances of degree 3, 4, 5, and 6. For a given degree, we present the reductions in order of priority.

## 5.8    Cubic Instances

Many formulae are not subject to any of the simplification rules above nor to large–component splitting. In this section we introduce further reductions so that for any formula $F$ of maximum degree at most 3 (which is to say, whose constraint graph has maximum degree at most 3), some reduction can be applied.

If $F$ has any vertex of degree strictly less than 3, we may apply the 0-, 1-, or 2-reductions above. Henceforth, then, we assume that $F$ is 3-regular (cubic).

The new reductions will generally be "atomic" in the sense that we will carry each through to its stated completion, not checking at any intermediate stage whether an earlier simplification or reduction rule can be applied.

We define

$$h_3 := a_3 - a_2 \tag{5.30}$$

to be the decrease of measure resulting from a half–edge reduction (reduction 5.5.6) on a vertex of degree 3.

### 5.8.1    3-cut

There is a 3-cut $X = \{x_1, x_2, x_3\}$ isolating a set $S$ of vertices, with $4 \leq |S| \leq 10$. Each cut vertex $x_i$ has at least 1 neighbor in $V \setminus \{S \cup X\}$ (otherwise $X$ without this vertex is a smaller cut), and without loss of generality we may assume that either each cut vertex has 2 neighbors in $V \setminus \{S \cup X\}$, or that $|S| = 10$. (If a cut vertex, say $x_1$, has just one neighbor $x_1' \in V \setminus \{S \cup X\}$, then $\{x_1', x_2, x_3\}$ is also a 3-cut, isolating the larger set $S \cup \{x_1\}$. Repeat until $|S| = 10$ or each cut vertex has two neighbors in $V \setminus \{S \cup X\}$.)

With reference to Figure 5.3, let $y_1, y_2, y_3 \in S$ be the respective neighbors of $x_1$, $x_2$, and $x_3$, and let $v_1$ and $v_2$ be the other neighbors of $x_1$. Note that $y_2 \neq y_3$, or we should instead apply a 2-cut reduction (reduction 5.5.10): cutting on $\{x_1, y_2\}$ isolates the set $S \setminus \{y_2\}$, and $3 \leq |S \setminus \{y_2\}| \leq 9$ satisfies the conditions of the 2-cut reduction.

We treat this case by branching on $x_1$, resulting in new instances $F_1$ and $F_2$. In each we apply a 2-cut on $\{y_2, y_3\}$ (not $\{x_2, x_3\}$!), creating a possibly–heavy edge $y_2 y_3$. We then 2-reduce on $y_2$ and $y_3$ in turn to create an edge $x_2 x_3$ which is heavy only if $x_2 y_2$ and $x_3 y_3$ were both heavy. If $|S| \leq 10$, the resulting instances satisfy

$$\mu(F_1), \mu(F_2) \leq \mu(F) - 5a_3 - 2h_3.$$

(Recall that for graphs of degree 3, $\mu$ and $\nu$ are identical.) The term $-5a_3$ accounts for the deletion of $x_1$ and $S$ (at least 5 vertices) with their incident half–edges. The term $-2h_3$ accounts for deletion of the "other halves" of the edges from $x_1$ to $V \setminus \{S \cup X\}$ and the degree decrease of their incident vertices (see definition (5.30)); we are using the fact that $v_1 \neq v_2$, and that $X$ is an independent set. There is no need for a term accounting for the deletion of the "other halves" of the edges on $x_2$ and $x_3$ and the addition of the new edge $x_2 x_3$: the new $x_2 x_3$ is heavy only if

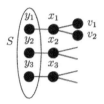

Figure 5.3: Illustration of a 3-cut, reduction 5.8.1

Figure 5.4: Illustration for reduction 5.8.2, on a vertex with independent neighbors

both half–edges were heavy, so this change in measure is $-\frac{1}{2}w(x_2y_2) - \frac{1}{2}w(x_3y_3) + w(x_2x_3) \leq 0$, and we are free to ignore it. (Since it may in fact be 0, there is also no gain including it.) Constraint (5.8′) of Lemma 5.1 is thus assured if

$$2^{-5a_3-2h_3} + 2^{-5a_3-2h_3} \leq 2^0 = 1,$$

that is it has a branching number of at most (see Section 2.6)

$$(5a_3 + 2h_3, 5a_3 + 2h_3). \tag{5.31}$$

By similar reasoning, if $|S| = 10$ the branching number is at most

$$(11a_3 + h_3, 11a_3 + h_3).$$

By (5.29) this constraint is bound to hold "for a sufficiently large value of 10" (and since $h_3 \leq a_3$, for 10 itself this constraint is dominated by (5.31)), so we will disregard it. ⊡

## 5.8.2 Vertex with Independent Neighbors

There is a vertex $u$ such that $N(u)$ is an independent set.

With reference to Figure 5.4, we reduce on $u$, fixing $\varphi(u)$ to 0 and 1 to generate new instances $F_0$ and $F_1$, each with constraint graph $G[V \setminus \{u\}]$.

Let $N^1 := N(u)$ and $N^2 := N^2(u)$. Let $q$ be the number of vertices in $N^1$ with a heavy edge to $N^2$, $k_0$ the number of vertices in $N^1$ subject to a super 2-reduction (deletion) in $F_0$, and $k_1$ the number subject to super 2-reduction in $F_1$. By Lemma 5.5, each $v \in N^1$ falls into at least one of these cases, so $q + k_0 + k_1 \geq 3$.

We will argue that $\mu(F) - \mu(F_i) \geq a_3 + 3h_3 + q(w_h - w_e) + 2k_ih_3$. Deletion of $u$ and reduction of the degree of each of its neighbors immediately reduces the measure by $a_3 + 3h_3$ (more if any edges incident to $u$ were heavy). In $F_i$, first 2-reduce on the $q$ vertices in $N^1$ with heavy edges (reducing the measure by a further $q(w_h - w_e)$) and on the $3 - q - k_i$ vertices subject to only plain 2-reductions (not increasing the measure). Note that each vertex in $N^2$ still has degree 3.

Finally, reduce out the $k_i$ vertices which are set constant by a super 2-reduction, by deleting their incident edges one by one. No vertex $v$ in $N^2$ has 3 neighbors in $N^1$: if it did there would remain only 3 other edges from $N^1$ to $N^2$, whence $|N^2| \leq 4$, $N^2 \setminus v$ would be a cut of size $\leq 3$ isolating $N^1 \cup \{u, v\}$, and we would have applied a cut reduction. Thus, deletion of each of the $2k_i$ edges in $N^1 \times N^2$ either reduces the degree of a vertex in $N^2$ from 3 to 2 (a good 1-reduction, reducing the measure by $h_3$), or creates a vertex of degree 1.

We wish to show that each degree-1 vertex in the graph $G' = G[V \setminus (\{u\} \cup N^1)]$ must also result in a good 1-reduction, giving the $2k_i h_3$ claimed. Note that $|N^2|$ must be 4, 5, or 6 (if it were smaller we would have applied a cut reduction instead). If $|N^2| = 6$ then every vertex in $N^2$ has degree 2 (in the graph $G'$) and there is nothing to prove. If $|N^2| = 5$ then at most one vertex in $N^2$ has degree 1, and Lemma 5.7 implies that it results in a good 1-reduction. If $|N^2| = 4$, every degree-1 vertex in $N^2$ also results in a good 1-reduction: If not, then by Lemma 5.7 a set $X$ of two or more vertices in $N^2$ lies in a small component of $G'$, in which case $N^2 \setminus X$ is a cut of size 2 or less in the original constraint graph $G$, isolating $\{u\} \cup N^1 \cup X$, and we would have applied a cut reduction instead.

Thus, $\mu(F) - \mu(F_i) \geq a_3 + 3h_3 + q(w_h - w_e) + 2k_i h_3$. By the Balance property on page 42, the worst cases come if $k_0 = 0$ and $k_1 = 3 - q$ (or vice–versa). Thus, the worst case branching numbers are

$$(\forall q \in \{0, 1, 2, 3\}) \quad \left( a_3 + 3h_3 + q(w_h - w_e), \right.$$
$$\left. a_3 + 3h_3 + q(w_h - w_e) + 2(3 - q)h_3 \right). \tag{5.32}$$

$\square$

### 5.8.3    One Edge in $G[N(u)]$

Given that we are in this case rather than Case 5.8.2, no vertex of $N(u)$ has an independent set as neighborhood. Let $N(u) = \{v_1, v_2, v_3\}$ and suppose without loss of generality that $v_2 v_3 \in E$. Let $N(v_1) = \{u, x_1, x_2\}$. Then, $x_1 x_2 \in E$. To avoid a 3-cut (Case 5.8.1), $|N^2(\{u, v_1\})| = 4$ (the 4 rightmost vertices depicted in Figure 5.5 are truly distinct).

After branching on $u$, in each of the two instances $F_0$ and $F_1$, first 2-reduce on $v_1$, then on $x_1$, then continue with 2-reductions (the base case), or super 2-reductions (if possible), on $v_2$ and $v_3$. In the base case this results in the deletion of all 5 of these vertices with their incident edges and the decrease of the degree of $x_2$ to 2, for a measure decrease of $5a_3 + h_3$ (vertex $x_2$ will be 2-reduced, which does not increase the measure; see 5.5.9).

If $v_2 v_3$ or $v_2 x_3$ is heavy, then there is an extra measure decrease of $w_h - w_e$ beyond that of

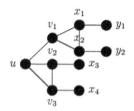

Figure 5.5: Illustration of reduction on a vertex with one edge in its neighborhood, Case 5.8.3

the base case, for a branching number of at most

$$(5a_3 + h_3 + w_h - w_e, 5a_3 + h_3 + w_h - w_e). \tag{5.33}$$

Otherwise, $v_2v_3$ and $v_2x_3$ are both light, and we may super 2-reduce on $v_2$ in either $F_0$ or $F_1$ (without loss of generality say $F_1$). This reduces the degree of $x_3$ from 3 to 2, and that of $v_3$ from 2 to 1, setting up a 1-reduction on $v_3$ that reduces the degree of $x_4$ from 3 to 2. This gives a branching number of at most

$$(5a_3 + h_3, 5a_3 + 3h_3). \tag{5.34}$$

⊡

There are no further cases for cubic graphs. If for a vertex $u$ there are 3 edges in $G[N(u)]$ then $N[u]$ is an isolated component (a complete graph $K_4$) and we apply component splitting. If there are 2 edges in $G[N(u)]$, then some $v \in N(u)$ (either of the vertices having a neighbor outside $\{u\} \cup N(u)$) has just 1 edge in $G[N(v)]$ and we are back to Case 5.8.3.

For results on cubic and other instances, we refer to Theorem 5.4, Table 5.2, and the discussion in Section 5.13.

### 5.8.4 Remark on Heavy Edges

If the original cubic instance is a pure 2-SAT formula, with no heavy edges, then (as we show momentarily) any heavy edges introduced by the procedure we have described can immediately be removed. Thus the "hybrid formula" concept gives no gain for cubic 2-SAT formulae, but expands the scope to cubic MAX 2-CSP, sacrifices nothing, and is useful for analyzing non cubic instances. We now show how heavy edges introduced into a cubic 2-SAT formula immediately disappear again.

In a graph with only light edges, the only two rules that create heavy edges are 2-reductions and 2-cuts (and other reductions that apply these). A 2-reduction on $v$ introduces a heavy edge only if $v$'s neighbors $x_1$ and $x_2$ were already joined by an edge. In that case, though, $x_1$ and $x_2$ have their degrees reduced to 2 (at most). If the remaining neighbors $y_1$ of $x_1$ and $y_2$ of $x_2$ are distinct, then 2-reducing on $x_1$ gives a light edge $x_2y_1$: the heavy edge $x_1x_2$ is gone. Otherwise, $y_1 = y_2$, and 2-reduction on $x_1$ followed by 1-reduction on $x_2$ deletes $x_1$ and $x_2$ and reduces the

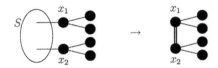

Figure 5.6: 2-cut rule creates a heavy edge

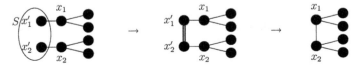

Figure 5.7: 2-cut rule avoids creating a heavy edge

degree of $y_2$ to 1, again leaving no heavy edge.

For a 2-cut on $x_1$ and $x_2$ isolating a set $S$, if there was an edge $x_1x_2$ then the cut reduction reduces the degrees of both $x_1$ and $x_2$ to 2, and, just as above, we may 2-reduce on $x_1$ to eliminate the heavy edge. If $x_1$ and $x_2$ are nonadjacent and $x_1$ has just 1 neighbor outside $S$, then again a follow–up 2-reduction on $x_1$ eliminates the heavy edge $x_1x_2$. Dismissing the symmetric case for $x_2$, all that remains is the case when $x_1$ and $x_2$ are not adjacent and each has 2 neighbors outside $S$, and thus just 1 neighbor in $S$; see Figure 5.6.

The $S$-neighbors $x_1'$ of $x_1$ and $x_2'$ of $x_2$ must be distinct, or else we would have applied a 1-cut reduction on $x_1'$. (This presumes that $|S \setminus \{x_1'\}| \geq 2$, but if it is 0 or 1, we would have 2-reduced on $x_1'$ or 1-reduced on its $S$-neighbor — either of which is really a special case of a 1-cut reduction.)

Given that $x_1' \neq x_2'$, apply a 2-cut reduction not on $x_1$ and $x_2$ but instead on $x_1'$ and $x_2'$. Following this with 2-reduction on $x_1'$ and $x_2'$ eliminates the heavy edge $x_1'x_2'$, giving a light edge $x_1x_2$ instead; see Figure 5.7.

### 5.8.5   Solving the Programs

Every weight constraint we introduce is of the form $\sum_i 2^{L_i} \leq 1$, where the sum is finite and each $L_i$ is some linear combination of weights. (Some constraints are simply of the form $L \leq 0$, but this can also be written as $2^L \leq 1$.) This standard form (along with the objective of minimizing $w_e$) can be provided, through an interface such as AMPL, as described in Section 2.4.

The convexity of the feasible region makes it relatively easy for a solver to return a provably optimal solution: convex programs are much easier to solve than general ones or even the quasiconvex programs like Eppstein's [Epp06]. IPOPT solves the nonlinear program for our general algorithm, to optimality, in a second or two on a typical laptop computer.

To ensure that our solutions are truly feasible, in the presence of finite numerical accuracy, we replace the "1" in the right–hand side of each constraint with $1 - \varepsilon$, fixing $\varepsilon := 10^{-6}$; this allows some margin for error. The values we show for the key parameters $w_e$ and $w_h$ are rounded up (pessimistically) from the higher–precision values returned by the solver, with the other parameter values rounded fairly. Ideally we would also verify, in an unlimited–accuracy tool such as Mathematica, that our rounded values satisfy the original "$\leq$ 1" constraints. Section 5.12 contains the final formulation of the mathematical program in AMPL.

## 5.9   Instances of Degree 4

Generalizing our earlier definition of $h_3$ (5.30), we define for any $d \geq 3$,

$$h_d := \min_{3 \leq i \leq d} \{a_i - a_{i-1}\}. \tag{5.35}$$

This is the minimum possible decrease of measure resulting from a half–edge reduction (reduction 5.5.6) on a vertex of degree $i$ with $3 \leq i \leq d$. We will find that such deletions always occur with the same sign in our nonlinear program — the larger $h_d$, the weaker each constraint is — and therefore the above definition can be expressed in our program by simple inequalities

$$(\forall 3 \leq i \leq d) \quad h_d \leq a_i - a_{i-1}. \tag{5.36}$$

We now consider a formula $F$ of (maximum) degree 4. The algorithm chooses a vertex $u$ of degree 4 with — if possible — at least one neighbor of degree 3. The algorithm sets $u$ to 0 and 1, simplifies each instance as much as possible (see Section 5.5), and recursively solves the resulting instances $F_0$ and $F_1$.

The instances $F_0$ and $F_1$ are either 4-regular, of degree at most 3, or nonregular. By the arguments presented in Section 5.3.6, the case where the degree of the graph decreases can be safely ignored (the measure decrease $C_4 - C_3$ can be made as large as necessary).

### 5.9.1   4-regular

If $F$ is 4-regular, first consider the case in which $F_0$ and $F_1$ are 4-regular. Since branching on $u$ decreases the degree of each vertex in $N(u)$, and none of our reduction rules increases the degree of a vertex, every vertex in $N(u)$ must have been removed from $F_0$ and $F_1$ by simplification rules.[2] This gives a branching number of at most

$$(5a_4, 5a_4). \tag{5.37}$$

---

[2]There is an important subtlety here: the reduced–degree vertices are eliminated, not merely split off into other components such that $F_i$ has a 4-regular component and a component of degree 3 (although such an example shares with 4-regularity the salient property that no degree-4 vertex has a degree-3 neighbor). By definition, the "4-regular case" we are considering at this point does not include such an $F_i$, but it is worth thinking about what happens to an $F_i$ which is not regular but has regular components. No component of $F_i$ is small (simplification 5.5.4 has been applied), so in the recursive solution of $F_i$, Algorithm max2csp immediately applies large–component splitting (reduction 5.7.1). This reduces $F_i$ to two connected instances, and is guaranteed to satisfy constraint (5.8') (the penalty for one instance's being 4-regular is more than offset by its being much smaller than $F_i$). Our machinery takes care of all of this automatically, but the example illustrates why some of the machinery is needed.

If neither $F_0$ nor $F_1$ is 4-regular, then $u$ is removed ($a_4$), the degree of its neighbors decreases ($4h_4$), and we obtain an additional gain because $F_0$ and $F_1$ are not regular ($R_4$). Thus, the branching number is at most

$$(a_4 + 4h_4 + R_4, a_4 + 4h_4 + R_4).  \tag{5.38}$$

If exactly one of $F_0$ and $F_1$ is 4-regular, we obtain a branching number of $(5a_4, a_4 + 4h_4 + R_4)$. By the dominance property on page 41, this constraint is weaker (no stronger) than (5.37) if $5a_4 \le a_4 + 4h_4 + R_4$, and weaker than (5.38) if $5a_4 > a_4 + 4h_4 + R_4$, so we may dispense with it.

## 5.9.2   4-nonregular

If $F$ is not 4-regular, we may assume that $u$ has at least one neighbor of degree 3. Let us denote by $p_i$ the number degree-$i$ neighbors of $u$. Thus, $1 \le p_3 \le 4$, and $p_3 + p_4 = 4$. Further, let us partition the set $P_3$ of degree-3 neighbors into those incident only to light edges, $P_3'$, and those incident to at least one heavy edge, $P_3''$. Define $p_3' := |P_3'|$ and $p_3'' := |P_3''|$ (so $p_3' + p_3'' = p_3$).

For each $F_i$ ($F_0$ and $F_1$), branching on $u$ removes $u$ (for a measure decrease of $a_4$, compared with $F$). If $F_i$ is not 4-regular, the degrees of the neighbors of $u$ all decrease ($\sum_{i=3}^4 p_i h_i$). If $F_i$ is regular ($-R_4$), all neighbors of $u$ must have been eliminated as well ($\sum_{i=3}^4 p_i a_i$).

We now argue about additional gains based on the values of $p_3'$ and $p_3''$, starting with the heavy edges incident on vertices in $P_3''$. Identify one heavy edge on each such vertex. If such an edge is between two vertices in $P_3''$ associate it with either one of them; otherwise associate it with its unique endpoint in $P_3''$. This gives a set of at least $\lceil p_3''/2 \rceil$ vertices in $P_3''$ each with a distinct associated heavy edge, which we may think of as oriented out of that vertex. If such an edge incident on $v \in P_3''$ is also incident on $u$ then it is deleted along with $u$, for an additional measure reduction of $w_h - w_e$ we credit to $v$. This leaves a set of "out" edges that may form paths or cycles. After deletion of $u$ all the vertices involved have degree 2, so any cycle is deleted as an isolated component, for a measure reduction of $w_h - w_e$ per vertex. Super 2-reducing on a vertex $v$ deletes its outgoing edge, which we credit to $v$, and possibly also an incoming heavy edge associated with a different $v' \in P_3''$, which we credit to $v'$. Finally, if $v$ is 2-reduced we consider its outgoing edge (not its other incident edge) to be contracted out along with $v$, crediting this to $v$ (and correctly resulting in a light edge if the other edge incident on $v$ was light, or a heavy one if it was heavy). This means that if the other edge incident to $v$ was a heavy edge out of a different $v' \in P_3''$, then $v'$ still has an associated outgoing heavy edge. In short, each of the $\lceil p_3''/2 \rceil$ vertices gets credited with the loss of a heavy edge, for an additional measure reduction of at least $\lceil p_3''/2 \rceil (w_h - w_e)$.

We say that we have a *good degree reduction* if the degree of a vertex of degree 3 or more decreases by 1: for graphs of degree 4 this decreases the measure by at least $h_4$. This measure decrease comes in addition to what we have accounted for so far, unless $F_i$ is regular and the degree reduction is on a vertex in $N(u)$ (since we have accounted for the deletion of those vertices, counting their degree reductions as well would be double counting). We will show that a certain number of additional-scoring degree reductions occur altogether, in $F_0$ and $F_1$ combined, as a function of $p_3'$.

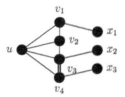

Figure 5.8: The case $p'_3 = 2$ may lead to just one good degree reduction outside $N[u]$. If both super 2-reductions on $v_1$ and $v_2$ occur in the same branch (say $F_1$), the degree of $x_1$ is reduced. The degrees of $v_3$ and $v_4$ become 2, so their edges are contracted eventually creating an edge $x_2x_3$, which does not change the degree of $x_2$ or $x_3$. The heavy edge $v_3v_4$ gives a bonus measure reduction of $w_h - w_e$ previously accounted for

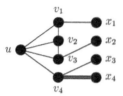

Figure 5.9: The case $p'_3 = 3$ ($P'_3 = \{v_1, v_2, v_3\}$) may lead to just two good degree reductions

If $p'_3 = 1$, super 2-reduction on the sole vertex in $P'_3$ is possible in at least one of $F_0$ or $F_1$ — without loss of generality say just $F_0$ — and reduces the degrees of at least two neighbors. If $F_0$ is nonregular this gives a gain of $2h_4$, while if $F_0$ is regular there may be no gain.

If $p'_3 = 2$, then again if either vertex is super 2-reduced in a nonregular branch there is a gain of at least $2h_4$. Otherwise, each vertex is super 2-reduced in a regular branch (both in one branch, or in two different branches, as the case may be). At least one of the vertices has at least one neighbor in $N^2 := N^2(u)$, or else $P_3 \setminus P'_3$ would be 2-cut. In whichever $F_i$ the degree of the neighbor is reduced, since $F_i$ is regular the neighbor must eventually be deleted, for a gain of at least $a_3$. So there is either a gain of $2h_4$ in a nonregular branch or a gain of $a_3$ in a regular branch. (We cannot hope to replace $a_3$ with $2a_3$: Figure 5.8 shows an example where indeed only one good degree reduction occurs outside $N[u]$.)

If $p'_3 = 3$, again either there is a gain of $2h_4$ in a nonregular branch, or each super 2-reduction occurs in a regular branch. The 3 vertices in $P'_3$ have at least 2 neighbors in $N^2$, or else these neighbors, along with $P_3 \setminus P'_3$, would form a cut of size 2 or smaller. Each of these neighbors has its degree reduced, and thus must get deleted from a regular $F_i$, for a gain of at least $2a_3$. So there is either a gain of $2h_4$ in a nonregular branch, or a gain of $2a_3$ altogether in one or two regular branches. (We cannot hope to claim $3h_4$ or $3a_3$, per the example in Figure 5.9.)

If $p'_3 = 4$, we claim that in the two branches together there are at least 4 good degree reductions on vertices in $N^2$ and $N^3(u)$. Each contributes a gain of at least $h_4$ if it is in a nonregular branch, $a_3$ in a regular branch. Each vertex in $N^2$ undergoes a good degree reduction in one branch or the other, so if $|N^2| \geq 4$ we are done. Since there can be no 2-cut, we may otherwise assume that $|N^2| = 3$. Since (in $F$) every vertex in $N(u)$ has degree 3, there is an even number of edges between $N(u)$ and $N^2$, thus there are at least 4 such edges. Since each vertex in $N^2$ has an edge from $N(u)$, there must be two such edges incident on one vertex $x_1 \in N^2$, and one edge each incident on the other vertices $x_2, x_3 \in N^2$. Again we guaranteed 4 good degree reductions unless $x_1$ has degree 3 *and* undergoes both of its reductions in one branch (so that degree 3 to 2 is a good reduction, but 2 to 1 is not). In that case, though, $x_1$ has degree 1, its remaining neighbor must be in $N^3(u)$ (otherwise $\{x_1, x_2\}$ is a 2-cut), and 1-reducing on $x_1$ gives a good degree reduction on that neighbor. So there is a total gain of $4h_4$ in a nonregular branch and $4a_3$ in a regular branch.

By convexity, the elementwise average of two pairs of branching numbers is a constraint dominated by one or the other, so it suffices to write down the extreme constraints, with all the gain from super 2-reductions given to a single nonregular or regular branch.

Before counting the super 2-reduction gains, if $F_i$ is nonregular the measure decrease $\mu(F) - \mu(F_i)$ is at least

$$\Delta_{\overline{r}}(p_3, p_3'', p_4) := a_4 + \sum_{i=3}^{4} p_i h_i + \left\lceil \tfrac{p_3''}{2} \right\rceil (w_h - w_e), \tag{5.39}$$

and if $F_i$ is 4-regular, at least

$$\Delta_r(p_3, p_3'', p_4) := a_4 + \sum_{i=3}^{4} p_i a_i + \left\lceil \tfrac{p_3''}{2} \right\rceil (w_h - w_e) - R_4. \tag{5.40}$$

The super 2-reductions give an additional gain, in a nonregular branch, of at least

$$g_{\overline{r}} := \left\lfloor \tfrac{p_3'+2}{3} \right\rfloor 2h_4, \tag{5.41}$$

and in a regular branch, at least

$$g_r := \left( \left\lfloor \tfrac{p_3'}{2} \right\rfloor + \left\lfloor \tfrac{p_3'}{3} \right\rfloor + \left\lfloor \tfrac{p_3'}{4} \right\rfloor \right) a_3, \tag{5.42}$$

where the tricky floor and ceiling expressions are just a way of writing an explicit expression convenient for passing to the nonlinear solver. The constraints arising from branching on a vertex of degree 4 with at least one neighbor of degree 3 are thus dominated by the following, taken over $p_3' + p_3'' + p_4 = 4$, with $p_4 \leq 3$ and $p_3 = p_3' + p_3''$:

$$(\Delta_{\overline{r}}, \Delta_{\overline{r}} + g_{\overline{r}}), \tag{5.43}$$

$$(\Delta_{\overline{r}}, \Delta_r + g_r), \tag{5.44}$$

$$(\Delta_r, \Delta_{\overline{r}} + g_{\overline{r}}), \tag{5.45}$$

$$(\Delta_r, \Delta_r + g_r). \tag{5.46}$$

## 5.10    Instances of Degree 5

This section considers formulae of maximum degree 5. As an overview, if there is a 3-cut isolating a set $S$ with 6 or more vertices and $S$ contains at least one vertex of degree 5, the algorithm branches on any vertex in the cut. Otherwise, the algorithm chooses a vertex $u$ of degree 5 with — if possible — at least one neighbor of degree at most 4, and branches on $u$ either as was done in the degree-4 case, or using clause–learning branching (see Lemma 5.6). We use clause learning when the neighbors of $u$ have high degrees, because clause learning sets many variables in $N(u)$, and this is most effective when the degrees are large (since $a_i \geq a_{i-1}$). We use normal branching when the neighbors have low degrees, because setting $u$ reduces their degrees, and this is effective when the degrees are small ($h_i \leq h_{i+1}$, with an additional bonus in super 2-reductions for a degree-3 variable). (This is also why we always prefer to branch on vertices of maximum degree with neighbors of low degree, and why the regular cases need special attention.)

### 5.10.1    3-cut

There is a 3-cut $C = \{x_1, x_2, x_3\}$ isolating a set $S$ of vertices such that $6 \leq |S| \leq 10$ and $S$ contains at least one vertex of degree 5. Branching on the cut vertex $x_1$ leaves constraint graphs where $\{x_2, x_3\}$ form a 2-cut. Thus $S \cup \{x_1\}$ are removed from both resulting instances ($a_5 + 6a_3$), a neighbor of $x_1$ outside $S \cup C$ has its degree reduced ($h_5$), a heavy edge $x_2 x_3$ appears (in the worst case) but at least 2 half-edges incident on $x_2$ and $x_3$ disappear ($-w_h + w_e$). Additionally, the resulting instances may become 5-regular ($-R_5$). So, the branching number is at most

$$(a_5 + 6a_3 + h_5 - w_h + w_e - R_5, \; a_5 + 6a_3 + h_5 - w_h + w_e - R_5). \qquad (5.47)$$

⊡

In light of reduction 5.10.1 we may henceforth assume that each degree-5 vertex $u$ has $|N^2(u)| \geq 4$.

### 5.10.2    5-regular

If every vertex has degree 5, the same analysis as for 4-regular instances (reduction 5.9.1, constraints (5.37) and (5.38)) gives a branching number which is at most one of the following:

$$(6a_5, 6a_5), \qquad (5.48)$$

$$(a_5 + 5h_5 + R_5, a_5 + 5h_5 + R_5). \qquad (5.49)$$

⊡

Otherwise, let $u$ be a degree-5 vertex with a minimum number of degree-5 neighbors, and as usual let $p_i$ be the number of degree-$i$ neighbors of $u$ (since the instance is not regular, $p_5 < 5$). Let $H := K_\delta(u$ is incident to a heavy edge). Depending on the values of $H$ and $p_i$ we will use either the usual 2-way branching (reduction 5.10.3) or clause–learning 3-way branching (reduction 5.10.4).

### 5.10.3 5-nonregular, 2-way Branching

In this case, $H = 1$ or $p_3 \geq 1$ or $p_5 \leq 2$.

We use the usual 2-way branching, setting $u$ to 0 and to 1, and simplifying to obtain $F_0$ and $F_1$. If $F_i$ is not regular, the measure decrease $\mu(F) - \mu(F_i)$ is at least $a_5 + \sum_{i=3}^{5} p_i h_i + H(w_h - w_e)$, and if $F_i$ is 5-regular, it is at least $a_5 + \sum_{i=3}^{5} p_i a_i + H(w_h - w_e) - R_5$. Thus if both branches are regular the branching number is at most

$$\left(a_5 + \textstyle\sum_{i=3}^{5} p_i a_i + H(w_h - w_e) - R_5, \; a_5 + \sum_{i=3}^{5} p_i a_i + H(w_h - w_e) - R_5\right), \tag{5.50}$$

and if one branch is regular and one nonregular, at most

$$\left(a_5 + \textstyle\sum_{i=3}^{5} p_i a_i + H(w_h - w_e) - R_5, \; a_5 + \sum_{i=3}^{5} p_i h_i + H(w_h - w_e)\right). \tag{5.51}$$

If both branches are nonregular, we use that if $p_3 \geq 1$, any degree-3 neighbor of $u$ either has a heavy edge not incident to $u$, giving an additional measure reduction of at least $w_h - w_e$, or in at least one branch may be super 2-reduced, for a measure reduction of at least $2h_5$. (The latter requires a justification we give explicitly, although Lemma 5.7 could be invoked. At the start of the first super 2-reduction, every vertex has degree 2 or more. Each of the two "legs" of the super 2-reduction propagates through a (possibly empty) chain of degree-2 vertices before terminating either in a good degree reduction or by meeting a vertex that was reduced to degree 1 by the other leg. In the latter case all the vertices involved had degree 2, thus were neighbors of $u$ originally of degree 3; also, there must have been at least three of them to form a cycle, and the remaining 2 or fewer vertices in $N(u)$ contradict the assumption that $F$ was simplified.) Thus, the branching number is at most

$$\left(a_5 + \textstyle\sum_{i=3}^{5} p_i h_i + H(w_h - w_e) + \mathsf{K}_\delta(p_3 \geq 1)2h_5, \; a_5 + \sum_{i=3}^{5} p_i h_i + H(w_h - w_e)\right) \quad \text{or} \tag{5.52}$$

$$\begin{aligned}
&\left(a_5 + \textstyle\sum_{i=3}^{5} p_i h_i + H(w_h - w_e) + \mathsf{K}_\delta(p_3 \geq 1)(w_h - w_e),\right. \\
&\left. a_5 + \sum_{i=3}^{5} p_i h_i + H(w_h - w_e) + \mathsf{K}_\delta(p_3 \geq 1)(w_h - w_e)\right).
\end{aligned} \tag{5.53}$$

$\boxed{\cdot}$

### 5.10.4 5-nonregular, Clause Learning

In this case, $H = 0$ and $p_3 = 0$ and $p_5 \in \{3, 4\}$.

Let $v$ be a degree 5 (degree 5 in $G$) neighbor of $u$ with a minimum number of degree-5 neighbors in $N^2 := N^2(u)$. The clause learning branching (see Lemma 5.6) will set $u$ in the first branch, $u$ and $v$ in the second branch, and all of $N[u]$ in the third branch. In each of the 3 branches, the resulting instance could become 5-regular or not.

In the *first branch*, the measure of the instance decreases by at least

$$\Delta_{51} := \min \begin{cases} a_5 + \sum_{i=4}^{5} p_i h_i & \text{(5-nonregular case),} \\ a_5 + \sum_{i=4}^{5} p_i a_i - R_5 & \text{(5-regular case).} \end{cases} \tag{5.54}$$

In the analysis of the second and third branches we distinguish between the case where $v$ has at most one neighbor of degree 5 in $N^2$, and the case where $v$ (and thus every degree-5 neighbor of $u$) has at least two neighbors of degree 5 in $N^2$.

In the *second branch*, if $v$ has at most one neighbor of degree 5 in $N^2$, the measure of the instance decreases by at least

$$\Delta_{52}^1 := \min \begin{cases} a_5 + \sum_{i=4}^{5} p_i h_i + a_4 + 3h_4 + h_5 & \text{(5-nonregular case),} \\ a_5 + \sum_{i=4}^{5} p_i a_i - R_5 & \text{(5-regular case).} \end{cases} \tag{5.55}$$

(The degree reductions $3h_4 + h_5$ from the nonregular case do not appear in the regular case because they may pertain to the same vertices as the deletions $\sum p_i a_i$.)

If $v$ has at least two neighbors of degree 5 in $N^2$, the measure decreases by at least

$$\Delta_{52}^2 := \min \begin{cases} a_5 + \sum_{i=4}^{5} p_i h_i + a_4 + 4h_5 & \text{(5-nonregular case),} \\ a_5 + \sum_{i=4}^{5} p_i a_i + 2a_5 - R_5 & \text{(5-regular case).} \end{cases} \tag{5.56}$$

In the *third branch*, first take the case where $v$ has at most one neighbor of degree 5 in $N^2$. Since $|N^2| \geq 4$, there are at least 4 good degree reductions on vertices in $N^2$. If the instance becomes regular, this implies a measure decrease of at least $4a_3$. If the instance remains nonregular, this is a measure reduction of at least $4h_5$, and we now show that if $p_5 = 4$ then there is a fifth good degree reduction. We argue this just as the 4-nonregular case (Section 5.9.2) with $p_3' = 4$; we could alternatively apply Lemma 5.7. If $|N^2| = 5$ the desired $5h_5$ is immediate. Otherwise, $|N^2| = 4$, and the number of edges between $N(u)$ and $N^2$ is at least 4, and odd (from $p_5 = 4$ and $p_4 = 1$, recalling that $p_3 = 0$), so at least 5. At least one edge incident on each vertex in $N^2$ gives a good degree reduction, and we fail to get a fifth such reduction only if the fifth edge is incident on a vertex $x \in N^2$ of degree 3, leaving it with degree 1. But in that case the remaining neighbor of $x$ must be in $N^3(u)$ (otherwise $N^2 \setminus x$ is a 3-cut, a contradiction by reduction 5.10.1), and 1-reducing $x$ gives the fifth good degree reduction. Thus the measure decreases by at least

$$\Delta_{53}^1 := \min \begin{cases} a_5 + \sum_{i=4}^{5} p_i a_i + 4h_5 + K_\delta(p_5 = 4)h_5 & \text{(5-nonregular case),} \\ a_5 + \sum_{i=4}^{5} p_i a_i + 4a_3 - R_5 & \text{(5-regular case).} \end{cases} \tag{5.57}$$

Otherwise, in the third branch, $v$ has at least two neighbors of degree 5 in $N^2$. For the regular case we simply note that each vertex in $N^2$ has its degree reduced and must be deleted, $N^2$ has at least four vertices of which at least two are of degree 5, for a measure reduction of at least $2a_5 + 2a_3$. We now address the nonregular case. Letting $P_5$ be the set of degree-5 vertices in $N(u)$ (so $|P_5| = p_5$), by definition of $v$ every vertex in $P_5$ has at least two degree-5 neighbors in $N^2$. Let $R \subseteq N^2$ be the set of degree-5 vertices in $N^2$ adjacent to $P_5$, and let $E_5 = E \cap (P_5 \times R)$ be the set of edges between $P_5$ and $R$. There is one last case distinction, according to the value of $p_5$. If $p_5 = 3$ there are at least 6 good degree reductions: $|E_5| = 6$, each vertex in $R$ has at most $|P_5| = 3$ incident edges from $E_5$, and thus each such incidence results in a good degree reduction (the vertex degree is reduced at most from 5 to 4 to 3 to 2). Here we have $6h_5$.

If $p_5 = 4$ we claim that the good degree reductions amount to at least $\min\{8h_5, 5h_5+h_4+h_3\}$. By default the 8 edges in $E_5$ all generate good degree reductions, with fewer only if some of the degree-5 vertices in $R$ have more than 3 incident edges from $E_5$. The "degree spectrum" on $R$ is thus a partition of 8 (the number of incident edges) into $|R|$ parts, where no part can be larger than $|P_4| = 4$. If the partition is $4 + 4$ this means two reductions that are not good ($2h_2$), but then this implies that $|R| = 2$, and the other two vertices in $N^2 \setminus R$ also have their degrees reduced, restoring the total of 8 good reductions. If the partition has exactly one 4, on a vertex $r \in R$, then just one of the 8 degree reductions is not good, and the 7 good reductions include those on $r$, thus giving a measure reduction of at least $5h_5 + h_4 + h_3$.

Considering the difference, which we will denote $g_{p_5=4}$, between these guaranteed measure decreases and the guarantee of $6h_5$ when $p_5 = 3$, we constrain

$$g_{p_5=4} \leq 8h_5 - 6h_5 = 2h_5, \tag{5.58}$$

$$g_{p_5=4} \leq (5h_5 + h_4 + h_3) - 6h_5 = -h_5 + h_4 + h_3. \tag{5.59}$$

and we obtain a measure reduction of at least

$$\Delta_{53}^2 := \min \begin{cases} a_5 + \sum_{i=4}^{5} p_i a_i + 6h_5 + \mathsf{K}_\delta(p_4 = 1)g_{p_5=4} & \text{(5-nonregular case)}, \\ a_5 + \sum_{i=4}^{5} p_i a_i + 2a_5 + 2a_3 - R_5 & \text{(5-regular case)}. \end{cases} \tag{5.60}$$

Wrapping up this reduction, the case that $v$ has at most 1 degree-5 neighbor in $N$, or at least two such neighbors, respectively impose the constraints (branching numbers)

$$(\Delta_{51}, \Delta_{52}^1, \Delta_{53}^1) \text{ and} \tag{5.61}$$

$$(\Delta_{51}, \Delta_{52}^2, \Delta_{53}^2). \tag{5.62}$$

$\square$

## 5.11   Instances of Degree 6

This section considers formulae of maximum degree 6. The algorithm chooses a vertex $u$ of degree 6 with — if possible — at least one neighbor of lower degree, and branches on $u$ by setting it to 0 and 1.

### 5.11.1    6-regular

If every vertex has degree 6, the same analysis as for regular instances of degree 4 gives a branching number which is at least one of the following:

$$(7a_6, 7a_6), \tag{5.63}$$
$$(a_6 + 6h_6 + R_6, a_6 + 6h_6 + R_6). \tag{5.64}$$

□

### 5.11.2    6-nonregular

Now, $u$ has at least one neighbor of degree at most 5.

It is straightforward that the branching number is at least as large as one of the following (only distinguishing if the instance becomes 6-regular or not):

$$\left(a_6 + \sum_{i=3}^{6} p_i h_i, \ a_6 + \sum_{i=3}^{6} p_i h_i\right), \tag{5.65}$$
$$\left(a_6 + \sum_{i=3}^{6} p_i a_i - R_6, \ a_6 + \sum_{i=3}^{6} p_i a_i - R_6\right). \tag{5.66}$$

□

## 5.12    Mathematical Program in AMPL

```
# maximum degree
param maxd integer >=3;
# fraction of non simple clauses
param p;
param margin;
set DEGREES := 0..maxd;
# weight for edges
var We >= 0;
# weight for degree reductions from degree at most i
var h {DEGREES} >= 0;
# vertex of degree i + i/2 surrounding half edges
var a {DEGREES};
# weight for heavy edges
var Wh;
# Regular weights
var R4 >= 0;                                              (5.14)
var R5 >= 0;                                              (5.14)
var R6 >= 0;                                              (5.14)
# additional degree reductions in the 3rd branch (nonregular)
```

```
# of the clause learning branching for p5=4 vs p5=3
var nonreg53;
# change in measure for the 3 branches
# 1st argument is the nb of deg-4 nbs of u
# 2nd argument distinguishes (if present) if v has at most 1 deg-5 nb
#                 in N^2 (1) or at least 2 (2)
set TWO := 1..2;
var f1 {TWO};
var f2 {TWO,TWO};
var f3 {TWO,TWO};
var D4r {0..4, 0..4};
var D4n {0..4, 0..4};
var g4r {0..4};
var g4n {0..4};


# analysis in terms of the number of edges
minimize Obj: (1-p)*We + p*Wh + 0*R4 + 0*R5 + 0*R6;

# Some things we know
subject to Known: a[0] = 0;                                              (5.24)

# Constrain W values non-positive
subject to Wnonpos {d in DEGREES : d>=1}:
  a[d] - d*We/2 <= 0 - margin;                                      (5.16)(5.12)

# a[] value positive
subject to MeasurePos {d in DEGREES : d>=1}:
  a[d] >= 0 + margin;                                                   (5.17)

# Intuition: weight for heavy edges >= weight for light edges
subject to HeavyEdge:
  We - Wh <= 0 - margin;                                                (5.19)

# collapse parallel edges
subject to parallel {d in DEGREES : d >= 3}:
   Wh - We - 2*a[d] + 2*a[d-1] <= 0 - margin;                          (5.22)

# decomposable edges
subject to Decomposable {d in DEGREES : d >= 1}:
   - a[d] + a[d-1] <= 0 - margin;                                      (5.25)

# constraints for the values of h[]
subject to hNotation {d in DEGREES, i in DEGREES : 3 <= i <= d}:
  h[d] - a[i] + a[i-1] <= 0 - margin;                              (5.30)(5.35)
```

```
#########################################
# constraints for cubic
#########################################


# 3-cut
subject to Cut3:
  2*2^(-5*a[3] - 2*h[3]) <= 1 - margin;                                    (5.31)

# Independent neighborhood
subject to Indep {q in 0..3}:
  2^(-a[3] - 3*h[3] -q*(Wh-We)) + 2^(-a[3] -3*h[3] - q*(Wh-We) - 2*(3-q)*h[3])
  <= 1 - margin;                                                           (5.32)

# One edge in neighborhood
subject to OneEdge1:
  2^(-5*a[3]-h[3]) + 2^(-5*a[3] -3*h[3]) <= 1 - margin;                    (5.34)
subject to OneEdge2:
  2^(-5*a[3] -h[3] -Wh +We) + 2^(-5*a[3] -h[3] -Wh +We) <= 1 - margin;     (5.33)


#########################################
# constraints for degree 4
#########################################


# 4-regular

# regular becomes nonregular
subject to Regular41:
  2* 2^(-a[4] - 4*h[4]-R4) <= 1 - margin;                                  (5.38)

# regular becomes regular
subject to Regular42:
  2* 2^(-5*a[4]) <= 1 - margin;                                           (5.37)

# 4 non-regular

subject to 4nonregularBase {p3p in 0..4, p3pp in 0..4, p4 in 0..3: p3p+p3pp+p4=4}:
  D4n[p3p,p3pp] = -a[4] -(p3p+p3pp)*h[3] -p4*h[4] -ceil(p3pp/2)*(Wh-We);   (5.39)

subject to 4regularBase {p3p in 0..4, p3pp in 0..4, p4 in 0..3: p3p+p3pp+p4=4}:
  D4r[p3p,p3pp] = -a[4] -(p3p+p3pp)*a[3] -p4*a[4] -ceil(p3pp/2)*(Wh-We) + R4;  (5.40)

subject to 4nonregularBonus {p3p in 0..4, p3pp in 0..4, p4 in 0..3: p3p+p3pp+p4=4}:
  g4n[p3p] = - floor((p3p+2)/3) * (2*h[4]);                                (5.41)

subject to 4regularBonus {p3p in 0..4, p3pp in 0..4, p4 in 0..3: p3p+p3pp+p4=4}:
```

```
g4r[p3p] = - (floor(p3p/2)+floor(p3p/3)+floor(p3p/4)) * a[3];          (5.42)


subject to Nonregular41 {p3p in 0..4, p3pp in 0..4, p4 in 0..3: p3p+p3pp+p4=4}:
  2^(D4n[p3p,p3pp]) + 2^(D4n[p3p,p3pp] + g4n[p3p])
  <= 1 - margin;                                                        (5.43)


subject to Nonregular42 {p3p in 0..4, p3pp in 0..4, p4 in 0..3: p3p+p3pp+p4=4}:
  2^(D4n[p3p,p3pp]) + 2^(D4r[p3p,p3pp] + g4r[p3p])
  <= 1 - margin;                                                        (5.44)


subject to Nonregular43 {p3p in 0..4, p3pp in 0..4, p4 in 0..3: p3p+p3pp+p4=4}:
  2^(D4r[p3p,p3pp]) + 2^(D4n[p3p,p3pp] + g4n[p3p])
  <= 1 - margin;                                                        (5.45)


subject to Nonregular44 {p3p in 0..4, p3pp in 0..4, p4 in 0..3: p3p+p3pp+p4=4}:
  2^(D4r[p3p,p3pp]) + 2^(D4r[p3p,p3pp] + g4r[p3p])
  <= 1 - margin;                                                        (5.46)


#######################################
# constraints for degree 5
#######################################

# 3-cut for degree 5
subject to Cut5_3:
  2* 2^(-a[5] - 6*a[3] + R5 +(Wh-We)) <= 1 - margin;                    (5.47)

# 5-regular

#  regular becomes nonregular
subject to Regular51:
  2* 2^(-a[5] - 5*h[5]-R5) <= 1 - margin;                              (5.48)

#  regular stays regular
subject to Regular52:
  2* 2^(-6*a[5]) <= 1 - margin;                                        (5.49)

# 5 non-regular

# clause learning

# first branch
subject to Cf1 {p4 in 1..2, p5 in 3..4: p4+p5=5}:
  f1[p4] >= -a[5]-p4*h[4]-p5*h[5];                                      (5.54)
subject to Cf1reg {p4 in 1..2, p5 in 3..4: p4+p5=5}:
  f1[p4] >= -a[5]-p4*a[4]-p5*a[5]+R5;                                   (5.54)
```

```
# second branch, v has at most 1 deg-5 neighbor in N^2
subject to Cf2a {p4 in 1..2, p5 in 3..4: p4+p5=5}:
    f2[p4,1] >= -a[5]-p4*h[4]-p5*h[5]-a[4]-3*h[4]-h[5];                    (5.55)
subject to Cf2areg {p4 in 1..2, p5 in 3..4: p4+p5=5}:
    f2[p4,1] >= -a[5]-p4*a[4]-p5*a[5]+R5;                                 (5.55)


# second branch, v (and all other deg-5 nbs of u) has >=2 deg-5 nbs in N^2
subject to Cf2b {p4 in 1..2, p5 in 3..4: p4+p5=5}:
    f2[p4,2] >= -a[5]-p4*h[4]-p5*h[5]-a[4]-4*h[5];                        (5.56)
subject to Cf2breg {p4 in 1..2, p5 in 3..4: p4+p5=5}:
    f2[p4,2] >= -a[5]-p4*a[4]-p5*a[5]-2*a[3]+R5;                          (5.56)


# additional degree reductions in the 3rd branch (nonregular) for p5=4 vs p5=3
subject to addDegRedNR53_1:
    nonreg53 <= 2*h[5];                                                   (5.58)
subject to addDegRedNR53_2:
    nonreg53 <= h[4]+h[3]-h[5];                                          (5.59)


# third branch, v has at most 1 deg-5 neighbor in N^2
subject to Cf3a {p4 in 1..2, p5 in 3..4: p4+p5=5}:
    f3[p4,1] >= -a[5]-p4*a[4]-p5*a[5]-(4+((4*p4+5*p5-5) mod 2))*h[5];     (5.57)
subject to Cf3areg {p4 in 1..2, p5 in 3..4: p4+p5=5}:
    f3[p4,1] >= -a[5]-p4*a[4]-p5*a[5]-4*a[3]+R5;                          (5.57)


# third branch, v (and all other deg-5 nbs of u) has >=2 deg-5 nbs in N^2
subject to Cf3b {p4 in 1..2, p5 in 3..4: p4+p5=5}:
    f3[p4,2] >= -a[5]-p4*a[4]-p5*a[5]-6*h[5]-floor(p5/4)*nonreg53;        (5.60)
subject to Cf3breg {p4 in 1..2, p5 in 3..4: p4+p5=5}:
    f3[p4,2] >= -a[5]-p4*a[4]-p5*a[5]-2*a[3]-2*a[5]+R5;                   (5.60)


# the clause learning splitting
subject to Nonregular5cl {p4 in 1..2, nb5 in 1..2}:
    2^(f1[p4]) + 2^(f2[p4,nb5]) + 2^(f3[p4,nb5]) <= 1 - margin;           (5.61)(5.62)


# 2-way splitting

# 2-way splitting, non-reg in both branches, if p3>0, then additional h-e
subject to Nonregular51a {p3 in 0..5, p4 in 0..5, p5 in 0..4, H in 0..1:
                          p3+p4+p5=5 and ((H=1) or (p5 < 3 or p3>0))}:
    2* 2^(-a[5] - p3*h[3] - p4*h[4] - p5*h[5] -H*(Wh-We) -ceil(p3/5)*(Wh-We))
    <= 1 - margin;                                                       (5.53)


# 2-way splitting, non-reg in both branches, if p3>0, then additional super-2
subject to Nonregular51b {p3 in 0..5, p4 in 0..5, p5 in 0..4, H in 0..1:
```

```
                        p3+p4+p5=5 and ((H=1) or (p5 < 3 or p3>0))}:
  2^(-a[5] - p3*h[3] - p4*h[4] - p5*h[5] -H*(Wh-We) -ceil(p3/5)*2*h[5])
+ 2^(-a[5] - p3*h[3] - p4*h[4] - p5*h[5] -H*(Wh-We))
<= 1 - margin;                                                          (5.52)
```

```
# 2-way splitting, becomes reg in both branches
subject to Nonregular52 {p3 in 0..5, p4 in 0..5, p5 in 0..4, H in 0..1:
                         p3+p4+p5=5 and ((H=1) or (p5 < 3 or p3>0))}:
  2* 2^(-a[5] - p3*a[3] - p4*a[4] - p5*a[5] -H*(Wh-We) + R5)
  <= 1 - margin;                                                        (5.50)
```

```
# 2-way splitting, becomes reg in 1 branch
subject to Nonregular52b {p3 in 0..5, p4 in 0..5, p5 in 0..4, H in 0..1:
                          p3+p4+p5=5 and ((H=1) or (p5 < 3 or p3>0))}:
  2^(-a[5] - p3*a[3] - p4*a[4] - p5*a[5] -H*(Wh-We) + R5)
+ 2^(-a[5] - p3*h[3] - p4*h[4] - p5*h[5] -H*(Wh-We))
<= 1 - margin;                                                          (5.50)
```

```
#######################################
# constraints for degree 6
#######################################
```

```
# 6-regular
```

```
# regular becomes nonregular
subject to Regular61:
  2* 2^(-a[6] - 6*h[6]-R6) <= 1 - margin;                               (5.64)
```

```
# regular stays regular
subject to Regular62:
  2* 2^(-7*a[6]) <= 1 - margin;                                         (5.63)
```

```
# 6 non-regular
```

```
# nonregular stays nonregular
subject to Nonregular61 {p3 in 0..6, p4 in 0..6, p5 in 0..6, p6 in 0..5:
                         p3+p4+p5+p6=6}:
  2* 2^(-a[6] - p6*h[6] - p5*h[5] - p4*h[4] - p3*h[3]) <= 1 - margin;   (5.65)
```

```
# nonregular becomes regular
subject to Nonregular62 {p3 in 0..6, p4 in 0..6, p5 in 0..6, p6 in 0..5:
                         p3+p4+p5+p6=6}:
  2* 2^(-a[6] - p6*a[6] - p5*a[5] - p4*a[4] - p3*a[3] +R6) <= 1 - margin; (5.66)
```

## 5.13   Tuning the Bounds

For any values of $w_e$ and $w_h$ satisfying the constraints we have set down, we have shown that any MAX 2-CSP instance $F$ is solved in time $\mathcal{O}^*\left(2^{|E|w_e+|H|w_h}\right)$.

For a given instance $F$, the running time bound is best for the feasible values of $w_e$ and $w_h$ which minimize $|E|w_e + |H|w_h$. As usual taking $|E| = (1-p)m$ and $|H| = pm$, this is equivalent to minimizing

$$(1-p)w_e + pw_h, \tag{5.67}$$

allowing us to obtain a 1-parameter family of running time bounds — pairs $(w_e, w_h)$ as a function of $p$ — tuned to a formula's fraction of conjunctive and general 2-clauses.

Reiterating, if a formula's "p" value is $p(F) = |H|/(|E| + |H|)$, and if minimizing (5.67) for a given $p$ gives a pair $(w_e, w_h)(p)$, then the optimal bound for formula $F$ is the one given by $(w_e, w_h)(p(F))$, but for *any* $(w_e, w_h)(p)$, the running time bound $\mathcal{O}^*\left(2^{|E|w_e+|H|w_h}\right)$ is valid for every formula $F$, even if $p \neq p(F)$. This is simply because every such pair $(w_e, w_h)$ is a feasible solution of the nonlinear program, even if it is not the optimal solution for the appropriate objective function.

For cubic instances, minimizing (5.67) with $p$ small gives $w_e \approx 0.10209$ and $w_h \approx 0.23127$, while minimizing with $p$ close to 1 gives $w_e = w_h = 1/6$ (the tight constraints are all linear, so the solution is rational), matching the best known polynomial space running time for general instances of MAX 2-CSP (see [SS07]). It appears that the first result is obtained for all $p \leq 1/2$ and the second for all $p > 1/2$.

For instances of degrees 4, 5, and 6 or more, the results of minimizing with various values of $p$ are shown in Table 5.2, and the most interesting of these is surely that of degree 6 or more (the general case). Here, taking $p$ small gives $w_e \approx 0.15820$ and $w_h \approx 0.31174$. For instances of MAX 2-SAT this gives a running time bound of $\mathcal{O}^*\left(2^{0.1582m}\right)$ or $\mathcal{O}^*\left(2^{m/6.321}\right)$, improving on the best bound previously known, giving the same bound for mixtures of OR and AND clauses, and giving nearly as good run times when a small fraction of arbitrary integer–weighted clauses are mixed in. We observe that any $p \geq 0.29$ leads to $w_e = w_h = 0.19$ (as for cubic case with $p > 1/2$, the tight constraints are linear, so the value is rational), matching the best known bound (for polynomial space algorithms) of $\mathcal{O}^*\left(2^{0.19m}\right)$ from [SS07]. Figure 5.10 shows the values of $w_e$, $w_h$, and the objective $(1-p)w_e + (p)w_h$, as a function of $p$. Numerically, the values $w_e$ and $w_h$ meet for some value of $p$ between 0.2899 and 0.29.

## 5.14   Conclusion

We have seen in this chapter a fast algorithm for MAX 2-SAT, MAX 2-CSP, and hybrid MAX 2-SAT/MAX 2-CSP instances. A rigorous analysis without much assumptions on the measure enabled us to establish a family of running time bounds, where the tight cases often depend on the parameter $p$ defining this family. To draw the graph of Figure 5.10, our convex program with close to 500 constraints was solved 3500 times: $p$ ranges from 0 to 0.35 and the values have

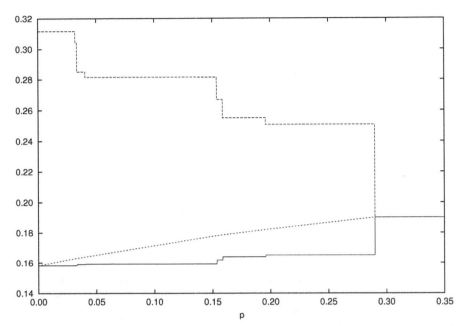

Figure 5.10: Plot of $w_e$ (solid), $w_h$ (dashed), and $(1-p)w_e + pw_h$ (dotted) versus the fraction $p$ of non–simple 2-clauses. The three values are equal (and exactly 0.19) for $p > 0.29$. Both $w_e$ and $w_h$ appear to be piecewise constant: the resolution of the graph is in $p$ increments of 0.0001, and all the small changes are meaningful

been computed by $p$ increments of 0.0001. To perform these computations was only a matter of minutes, which shows the power of the convex programming method to optimize a measure.

If we only look at MAX 2-SAT, our main improvement comes from the fact that we use powerful reductions taking us outside the class of MAX 2-SAT instances, account for these by an appropriate measure and carefully analyze the algorithm. The algorithm we presented sometimes splits off a part of the graph when it finds a 2-cut in the graph by introducing a CSP–clause between the two vertices of the 2-cut. CSP clauses are a generalization concerning the type of clauses. It would be interesting to know if a generalization concerning the size of the clauses would be useful. One could, for example, split off a part of the graph that is separated from the rest of the graph by a 3-cut and introduce a CSP clause involving the 3 vertices of the 3-cut.

# Chapter 6

# Treewidth Bounds

Equations are more important to me, because
politics is for the present, but an equation is
something for eternity.

*Albert Einstein*

Chapters 7 and 8 present algorithms combining branching algorithms and tree- or pathwidth based algorithms in different ways. During the execution of the algorithms, tree decompositions need to be computed. The smaller the width of these decompositions, the faster the dynamic programming algorithms using these decompositions perform. In this section, we discuss bounds on the tree decompositions of graphs according to the degrees of their vertices. These bounds are proved constructively and imply polynomial time procedures for finding decompositions whose widths do not exceed the respective bounds.

**Definition 6.1.** A *tree decomposition* of a graph $G = (V, E)$ is a pair $(\{X_i : i \in I\}, T)$ where each $X_i$, $i \in I$, is a subset of $V$, called a *bag* and $T$ is a tree with elements of $I$ as nodes such that

1. $\bigcup_{i \in I} X_i = V$,

2. for all $uv \in E$, there exists $i \in I$ such that $\{u, v\} \subseteq X_i$, and

3. for all $i, j, k \in I$, if $j$ is on the path from $i$ to $k$ in $T$ then $X_i \cap X_k \subseteq X_j$.

The *width* of a tree decomposition is $\max_{i \in I} |X_i| - 1$. The *treewidth* of a graph $G$ is the minimum width over all its tree decompositions and it is denoted $\mathbf{tw}(G)$. A tree decomposition $(\{X_i : i \in I\}, T)$ is a *path decomposition* if $T$ is a path. The *pathwidth* of a graph $G$ is the minimum width over all its path decompositions and it is denoted $\mathbf{pw}(G)$. A tree decomposition (respectively, a path decomposition) is called *optimal* if its width is $\mathbf{tw}(G)$ (respectively $\mathbf{pw}(G)$).

Algorithms using tree decompositions often need nice tree decompositions.

**Definition 6.2** (Nice tree decomposition). A *nice tree decomposition* $(\{X_i : i \in I\}, T)$ is a tree decomposition satisfying the following properties:

1. every node of $T$ has at most two children;

2. if a node $i$ has two children $j$ and $k$, then $X_i = X_j = X_k$ ($i$ is called a Join Node);

3. if a node $i$ has one child $j$, then either

   (a) $|X_i| = |X_j| + 1$ and $X_j \subset X_i$ ($i$ is called an Insert Node), or
   (b) $|X_i| = |X_j| - 1$ and $X_i \subset X_j$ ($i$ is called a Forget Node).

That a tree decomposition of a graph $G$ can be transformed in linear time into a nice tree decomposition of the same width and with at most 4 times as many bags has been shown by Kloks.

**Theorem 6.3** ([Klo94]). *For a constant $k$, given a tree decomposition of a graph $G$ of width $k$ and $N$ bags, one can find a nice tree decomposition of $G$ of width $k$ with at most $4N$ bags in $O(n)$ time, where $n$ is the number of vertices of $G$.*

The graph parameters treewidth and pathwidth were introduced by Robertson and Seymour in their seminal work on graph minors [RS83, RS86]. They play nowadays a central role in algorithmic graph theory as many $\mathcal{NP}$-hard problems become polynomial time solvable on graphs of small treewidth. For a survey on treewidth based (sub)exponential time algorithms we refer to [FGK05].

## 6.1   Bounds on the Pathwidth of Sparse Graphs

In this section we develop several upper bounds on the pathwidth of sparse graphs. We need the following known bound on the pathwidth of graphs with maximum degree 3 to prove the two lemmata of this section. It has been proved by Fomin and Høie based on the work of Monien and Preis on the bisection width of 3–regular graphs [MP06].

**Theorem 6.4** ([FH06]). *For any $\varepsilon > 0$, there exists an integer $n_\varepsilon$ such that for every graph $G$ with $n > n_\varepsilon$ vertices and maximum degree at most 3, $\mathbf{pw}(G) \leq (1/6 + \varepsilon)n$. Moreover, a path decomposition of the corresponding width can be constructed in polynomial time.*

Using Theorem 6.4 we prove the following bound for general graphs.

**Lemma 6.5.** *For any $\varepsilon > 0$, there exists an integer $n_\varepsilon$ such that for every graph $G$ with $n > n_\varepsilon$ vertices,*

$$\mathbf{pw}(G) \leq \frac{1}{6}n_3 + \frac{1}{3}n_4 + \frac{13}{30}n_5 + \frac{23}{45}n_6 + n_{\geq 7} + \varepsilon n,$$

*where $n_i$ is the number of vertices of degree $i$ in $G$ for any $i \in \{3, \dots, 6\}$ and $n_{\geq 7}$ is the number of vertices of degree at least 7. Moreover, a path decomposition of the corresponding width can be constructed in polynomial time.*

*Proof.* Let $G = (V, E)$ be a graph on $n$ vertices. It is well known (see for example [Bod98]) that if the treewidth of a graph is at least 2, then contracting edges incident to vertices of degree 1 and 2 does not change the treewidth of a graph and thus increases its pathwidth by at most a logarithmic factor, as shown in [KS93]. So we assume that $G$ has no vertices of degree 1 and 2 (otherwise we contract the corresponding edges). Furthermore, adding loops and duplicating edges does not increase the pathwidth of a graph, so we may at all times assume that the graph is simple.

First, we prove the lemma for the special case where the maximum degree of $G$ is at most 4 by induction on the number $n_4$ of vertices of degree 4 in $G$. If $n_4 = 0$, then $\Delta(G) \leq 3$ and we apply Theorem 6.4. Let us assume that the lemma holds for graphs with at most $n_4 - 1 \geq 0$ and prove it for graphs with $n_4$ vertices of degree 4. Let $v \in V$ be a vertex of degree 4. Let $i \in \{0, \ldots, 4\}$ be the number of degree 3 neighbors of $v$. As every neighbor of $v$ has degree at least 3, $v$ has $4 - i$ neighbors of degree 4. Adding $v$ to every bag of the tree decomposition increases the width of this tree decomposition by 1. Thus,

$$
\begin{aligned}
\mathbf{pw}(G) &\leq \mathbf{pw}(G \backslash v) + 1 \\
&\leq \frac{n_3 - i + (4 - i)}{6} + \frac{n_4 - 1 - (4 - i)}{3} + \varepsilon(n - 1) + 1 \\
&\leq \frac{n_3}{6} + \frac{n_4}{3} + \varepsilon n \ .
\end{aligned}
$$

Now, suppose that the maximum degree of $G$ is at most 5. The case where the graph or some of its connected components is 5-regular needs special consideration. Note that the pathwidth of a graph equals the maximum pathwidth of all its connected components, so it is sufficient to prove the bound for the connected component of a graph that has largest pathwidth (or for all connected components separately). 5-regular connected components may occur in the following situations:

(a) the input graph is 5-regular or has 5-regular connected components,

(b) the removal of a vertex of degree at least 6 led to a graph with one or more 5-regular components,

(c) the removal of a vertex of degree 5 produced 5-regular connected components by splitting off all vertices of degree at most 4 into different connected components, and

(d) the removal of a vertex of degree 5 produced a 5-regular graph (by contracting edges incident to vertices of degree 1 and 2).

We have already proved the base case where $n_5 = 0$. Let us assume that the lemma holds for all graphs with at most $n_5 - 1$ vertices of degree 5, no vertices of degree at least 6 and at least one vertex of degree at most 4. The case when the graph is 5-regular is considered later.

Let $v$ be a vertex of degree 5 with at least one neighbor of degree at most 4. Let $G'$ be the connected component of largest pathwidth of the graph obtained from $G - v$ by contracting edges incident to vertices of degree 1 and 2. Let us first assume that $G'$ is not 5-regular. It is clear that $\mathbf{pw}(G) \leq \mathbf{pw}(G \setminus v) + 1$. For $j \in \{3, \ldots, 5\}$ we denote by $m_j$ the number of degree $j$ neighbors of $v$. By the induction assumption,

$$\mathbf{pw}(G) \leq \mathbf{pw}(G \setminus v) + 1$$
$$\leq \frac{n_3 - m_3 + m_4}{6} + \frac{n_4 - m_4 + m_5}{3} + \frac{13}{30}(n_5 - 1 - m_5) + 1 + \varepsilon(n - 1).$$

For all possible values of $(m_3, m_4, m_5)$, we have that

$$\frac{-m_3 + m_4}{6} + \frac{-m_4 + m_5}{3} + \frac{13}{30}(-1 - m_5) + 1 \leq 0.$$

(Equality is obtained when $(m_3, m_4, m_5) = (0, 1, 4)$ which corresponds to the case when $v$ has four neighbors of degree 5 and one of degree 4.) Thus,

$$\mathbf{pw}(G) \leq \frac{n_3}{6} + \frac{n_4}{3} + \frac{13}{30}n_5 + \varepsilon n.$$

If the graph $G'$ is 5-regular, then all neighbors of $v$ in $G$ are removed either (d) by contracting edges incident to vertices of degree 1 and 2 or (c) by splitting off vertices of degree at most 4 into a different connected component. In the worst case, all neighbors of $v$ are of degree 3 in this case. Let $u$ be a vertex of degree 5 in $G'$. Since $G' \setminus u$ is not 5-regular and $\mathbf{pw}(G) \leq \mathbf{pw}(G' \setminus u) + 2$, we have that

$$\mathbf{pw}(G) \leq \mathbf{pw}(G' \setminus u) + 2$$
$$\leq 2 + \frac{n_3 - 5}{6} + \frac{n_4 + 5}{3} + \frac{13}{30}(n_5 - 7) + \varepsilon(n - 2)$$
$$< \frac{n_3}{6} + \frac{n_4}{3} + \frac{13}{30}n_5 + \varepsilon n.$$

Thus the lemma holds for all non 5-regular graphs. Since the removal of one vertex (for cases (a) and (b)) changes the pathwidth by an additive factor of at most 1, for sufficiently large $n$ this additive factor is dominated by $\varepsilon n$, and we conclude that the lemma holds for 5-regular graphs as well.

Using similar arguments one can proceed with the vertices of degree 6 (we skip the proof here). The critical case here is when a vertex of degree 6 has 5 neighbors of degree 6 and one neighbor of degree 5.

For vertices of degree at least 7 we just use the fact that adding a vertex to a graph can increase its pathwidth by at most one.                                                                      □

More accurate bounds for vertices of degree at least 7 can be obtained by a computer program going through all possible cases. The obtained values are reported in Table 6.1.

As a corollary, we get the following bound which was proved in [KMRR09] and [SS07].

**Corollary 6.6.** *For any $\varepsilon > 0$, there exists an integer $n_\varepsilon$ such that for every graph $G$ with $n > n_\varepsilon$ vertices and $m$ edges, $\mathbf{pw}(G) \leq 13m/75 + \varepsilon n$.*

| $d$ | $\beta_d$ | $d$ | $\beta_d$ | $d$ | $\beta_d$ |
|-----|-----------|-----|-----------|-----|-----------|
| 3   | 0.1667    | 8   | 0.6163    | 13  | 0.7514    |
| 4   | 0.3334    | 9   | 0.6538    | 14  | 0.7678    |
| 5   | 0.4334    | 10  | 0.6847    | 15  | 0.7822    |
| 6   | 0.5112    | 11  | 0.7105    | 16  | 0.7949    |
| 7   | 0.5699    | 12  | 0.7325    | 17  | 0.8062    |

Table 6.1: Numerically obtained constants $\beta_d$, $3 \leq d \leq 17$, such that for any $\varepsilon > 0$, there exists an integer $n_\varepsilon$ such that for every graph $G$ with $n > n_\varepsilon$ vertices, $\mathbf{pw}(G) \leq \sum_{d=3}^{17} \beta_d n_d + n_{\geq 18} + \varepsilon n$, where $n_i$ is the number of vertices of degree $i$ in $G$ for any $i \in \{3, \ldots, 17\}$ and $n_{\geq 18}$ is the number of vertices of degree at least 18

*Proof.* First, suppose $G$ has maximum degree at most 5. Then every edge in $G$ contributes at most

$$\max_{3 \leq d \leq 5} \left\{ \frac{2\beta_d}{d} \right\}$$

to the pathwidth of $G$, where $\beta_3 = 1/6, \beta_4 = 1/3, \beta_5 = 13/30$ are the values from Lemma 6.5. The maximum is obtained for $d = 5$ and is $13/75$. Thus, the result follows for graphs of maximum degree at most 5.

Finally, if $G$ has a vertex $v$ of degree at least 6, then we use induction on the number of vertices of degree at least 6. The base case has already been proved and the inductive argument is as follows:

$$\mathbf{pw}(G) \leq \mathbf{pw}(G \setminus v) + 1 \leq 13(m - 6)/75 + 1 < 13m/75.$$

$\square$

The following result bounds the pathwidth of a graph in terms of both the number of vertices and the number of edges and is very useful when we only have information about the average degree of a graph with at most $2n$ edges.

**Lemma 6.7.** *For any $\varepsilon > 0$, there exists an integer $n_\varepsilon$ such that for every connected graph $G$ with $n > n_\varepsilon$ vertices and $m = \beta n$ edges, $1 \leq \beta \leq 2$, the pathwidth of $G$ is at most $(m-n)/3 + \varepsilon n$. Moreover, a path decomposition of the corresponding width can be constructed in polynomial time.*

*Proof.* First we show the result assuming that the maximum degree $\Delta(G)$ of the graph is bounded by 3, then we extend this result to the general case.

Let $n_2$ be the number of vertices of degree 2 in $G$ and $n_3$ be the number of vertices of degree 3 in $G$. Since the contraction of an edge incident to a vertex of degree one does not change the treewidth of a graph, we assume that $n_2 = n - n_3$. Thus $n_2 + \frac{3}{2}n_3 = \beta n$. Since $n_3 = 2(\beta - 1)n$,

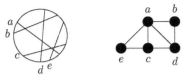

Figure 6.1: A circle model and the corresponding circle graph

by Lemma 6.5 we have that

$$\mathbf{pw}(G) \leq \frac{1}{6}(2\beta - 2)n + \varepsilon n$$
$$= \frac{\beta - 1}{3}n + \varepsilon n = \frac{m - n}{3} + \varepsilon n.$$

Now we extend the result without any assumptions on the degrees of the vertices of $G$. We show this by induction on $n_{\geq 4}$, the number of vertices of degree at least 4. We have already shown that the lemma holds if $n_{\geq 4} = 0$. Let us assume that for every $\varepsilon > 0$ there exists $n_\varepsilon$ such that for every graph with at least $n_\varepsilon$ vertices and at most $n_{\geq 4} - 1$ vertices of degree at least 4 the lemma holds. Let $v \in V$ be a vertex of degree at least 4. Observe that $G \setminus v$ has $n - 1$ vertices and at most $m - 4 \leq \beta(n - 1)$ edges. Now we have

$$\mathbf{pw}(G) \leq \mathbf{pw}(G \setminus v) + 1 \leq \frac{(m - 4) - (n - 1)}{3} + 1 + \varepsilon(n - 1)$$
$$\leq \frac{m - n}{3} + \varepsilon n.$$

$\square$

## 6.2   Bound on the Treewidth of Circle Graphs

In this section, we present a bound on the treewidth of circle graphs in terms of their maximum degree: $\mathbf{tw}(G) \leq 4\Delta(G)$ for every circle graph $G$.

**Definition 6.8.** A *circle graph* is an intersection graph of chords in a circle. More precisely, $G$ is a circle graph if there is a circle with a collection of chords, such that one can associate in a one-to-one manner a chord to each vertex of $G$ such that two vertices are adjacent in $G$ if and only if the corresponding chords have a nonempty intersection. The circle and all the chords are called a *circle model* of the graph.

We refer to Figure 6.1 for an example of a circle graph and its corresponding circle model.

Our approach is based on the fundamental ideas of Kloks' algorithm to compute the treewidth of circle graphs [Klo96]. We start with a brief summary of this algorithm. Consider the circle model of a circle graph $G$. Go around the circle and place a new point (a so-called *scanpoint*) between every two consecutive end points of chords. The treewidth of a

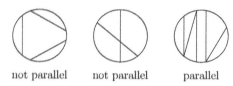

not parallel        not parallel         parallel

Figure 6.2: Examples of parallel and non parallel sets of chords

circle graph can be computed by considering all possible triangulations of the polygon $\mathcal{P}$ formed by the convex hull of these scanpoints. The weight of a triangle in this triangulation is the number of chords in the circle model that cross this triangle. The weight of triangulation $\mathcal{T}$ is the maximum weight of a triangle in $\mathcal{T}$. The treewidth of the graph is the minimum weight minus one over all triangulations $\mathcal{T}$ of $\mathcal{P}$. To find an optimal tree decomposition of $G$, the algorithm in [Klo96] uses dynamic programming to compute a minimum weight triangulation of $\mathcal{P}$.

**Theorem 6.9** ([Klo96]). *There exists an $O(n^3)$ algorithm to compute the treewidth of circle graphs, that also computes an optimal tree decomposition.*

We rely on the following technical definitions in our construction of a tree decomposition of width at most $4\Delta(G)$ for each circle graph $G$. The construction will be given in the proof of Theorem 6.13.

**Definition 6.10.** A *scanline* $\tilde{s} = \langle \tilde{a}, \tilde{b} \rangle$ is a chord connecting two scanpoints $\tilde{a}$ and $\tilde{b}$.

To avoid confusion, we call *vertex chords* the chords of the circle model that represent the vertices of the corresponding circle graph. Scanlines are chords as defined above and the general term *chord* refers to both scanlines and vertex chords. To emphasize the difference between scanlines and vertex chords we use different notations: A vertex chord $v$ connecting two end points $c$ and $d$ in the circle model of the graph is denoted $v = [c, d]$. This notation is also used if we consider chords in general. We adapt the standard convention that two vertex chords never intersect on the circle. Moreover, we say that two chords with empty intersection or intersecting in exactly one point on the circle (scanpoint) are *non-crossing*.

**Definition 6.11.** Let $c_1$ and $c_2$ be two non-crossing chords. A chord $c$ is *between* $c_1$ and $c_2$ if every path from an end point of $c_1$ to an end point of $c_2$ along the circle passes through an end point of $c$.

**Definition 6.12.** A set $C$ of chords is *parallel* if and only if

(i) the chords of $C$ are non-crossing, and

(ii) if $|C| > 2$, then for every subset of three chords in $C$, one of these chords is between the other two.

Algorithm **TriangCircle(circle model of a graph $G$)**
**Input**   : A circle model of a graph $G$.
**Output**: A triangulation of weight at most $4\Delta(G)$ of the polygon defined by the
                  scanpoints of this circle model.

Choose any vertex chord $v$ in the circle model of $G$
$S \leftarrow$ **ScanChord**$(\emptyset, v)$
**return ParaCuts**$(S)$

Figure 6.3: Algorithm **TriangCircle** computing a triangulation of weight at most $4\Delta(G)$
of any circle graph $G$

A set $S$ of scanlines is *maximal parallel* if there exists no vertex chord $v$ such that $S \cup \{v\}$ is parallel. Given a maximal parallel set of scanlines $S$, consider the maximal size subpolygons of $\mathcal{P}$ that do not properly intersect any scanline of $S$ (but there may be scanlines of $S$ on their boundaries). For such a subpolygon of $\mathcal{P}$, either one or two edges are scanlines of $S$. We say that these polygons are *delimited* by one or two scanlines of $S$ and we call *outer polygon* $\mathcal{P}_{\tilde{s}}$ *with respect to* $S$ such a polygon delimited by one scanline $\tilde{s} \in S$ and *inner polygon* $\mathcal{P}_{\tilde{s}_1, \tilde{s}_2}$ *with respect to* $S$ such a polygon that is delimited by two scanlines $\tilde{s}_1, \tilde{s}_2 \in S$ and contains at least one scanpoint (otherwise, it is already triangulated). The inner and outer polygons are defined with respect to a maximal parallel set of scanlines $S$, but we allow ourselves to not state this set of scanlines explicitly if it is clear from the context.

The following theorem shows that the treewidth $\mathbf{tw}(G)$ of every circle graph $G$ can be upper bounded by a linear function of the maximum degree $\Delta(G)$ of the graph.

The idea for the proof is to construct an algorithm that computes a triangulation of $\mathcal{P}$ (the triangulation is not necessarily optimal) and to prove that each triangle of this triangulation has weight at most $4\Delta(G)$. Before presenting the algorithm in detail, let us mention some of its major ideas. The algorithm separates $\mathcal{P}$ into "slices" by scanning some appropriately chosen vertex chords in the circle model of the graph, where a vertex chord $v$ is scanned by adding two sharp triangles to the partly constructed triangulation: two scanlines parallel to $v$ and one scanline crossing $v$ to form two triangles. The slices are made thinner and thinner by adding scanlines to the partly constructed triangulation until no slice can be cut into a pair of slices by scanning a vertex chord any more, and this procedure gives a maximal parallel set of scanlines. When triangulating the "middle part" of any slice (a convex polygon formed by the endpoints of the two delimiting scanlines and possibly other scanpoints on one side of the slice), we use the property that no vertex chord is parallel to the two scanlines delimiting the slice to show that the algorithm will not create triangles with a weight exceeding $4\Delta(G)$. The borders of the slices are triangulated recursively by first separating them into slices (in the perpendicular orientation of the previous slices) by scanning some chords and processing the resulting slices similarly.

The most interesting procedure of our algorithm is **TriangInner**, which is also crucial for our upper bound $4\Delta(G)$.

**Theorem 6.13.** *For every circle graph $G$, $\mathbf{tw}(G) \leq 4\Delta(G)$.*

---

Procedure **ScanChord**$(S, v = [a, b])$
**Input** : A set of scanlines $S$ and a vertex chord $v = [a, b]$ such that no scanline of $S$
crosses $v$.
**Output**: A set of scanlines triangulating the polygon defined by the neighboring
scanpoints of the end points of $v$.

Let $\tilde{c}$ and $\tilde{c}'$ (respectively $\tilde{d}$ and $\tilde{d}'$) be the two scanpoints closest to $a$ (respectively $b$)
such that the order of the points on the circle is $\tilde{c}, a, \tilde{c}', \tilde{d}', b, \tilde{d}$
Let $\tilde{s}_1 := \langle \tilde{c}, \tilde{d} \rangle$, $\tilde{s}_2 := \langle \tilde{c}', \tilde{d}' \rangle$ and $\tilde{s}_3 := \langle \tilde{c}, \tilde{d}' \rangle$
**if** $\tilde{c} = \tilde{d}$ (or $\tilde{c}' = \tilde{d}'$) **then**
$\quad \lfloor \; X \leftarrow \{\tilde{s}_2\} \; (\text{or } \{\tilde{s}_1\})$
**else**
$\quad \lfloor \; X \leftarrow \{\tilde{s}_1, \tilde{s}_2, \tilde{s}_3\}$
**return** $X$

Figure 6.4: Procedure **ScanChord** producing a set of scanlines triangulating the polygon
defined by the neighboring scanpoints of a vertex chord

---

Figure 6.5: Illustration of **ScanChord**$(S, v = [a, b])$

*Proof.* The theorem clearly holds for edgeless graphs. Let $G$ be a circle graph with at least one edge and $\mathcal{P}$ be the polygon as previously described. We construct a triangulation of $\mathcal{P}$ such that every triangle has weight at most $4\Delta$, that is it intersects at most $4\Delta$ vertex chords, and therefore the corresponding tree decomposition has width at most $4\Delta - 1$.

Notice that by the definition of a circle graph, every vertex chord intersects at most $\Delta$ other vertex chords. The triangulation of the polygon $\mathcal{P}$ is obtained by constructing the corresponding set of scanlines $S$ which is explained by the following procedures. Along with the description of our algorithm, we also analyze the number of vertex chords that cross each triangle and show that it is at most $4\Delta$.

We say that a procedure is valid if it does not create triangles with weight higher than $4\Delta$ and if it does not create crossing scanlines.

The validity of Algorithm **TriangCircle** depends on the validity of the procedures **Scan-Chord** and **ParaCuts**. Note that, initially, no scanline crosses $v$, which is a condition for **ScanChord**. Moreover **ScanChord** produces a parallel set of scanlines, which is a condition for **ParaCuts**.

The procedure **ScanChord** returns a set $X$ of one or three scanlines. They form at most two triangles: $\tilde{c}, \tilde{d}, \tilde{d}'$ and $\tilde{c}, \tilde{d}', \tilde{c}'$. Each of them intersects at most $\Delta + 1$ vertex chords: $v$ and the vertex chords crossing $v$. Furthermore, at most $\Delta$ vertex chords cross $\tilde{s}_1$ and $\tilde{s}_2$, precisely the vertex chords that cross $v$. The scanlines of $X$ do not intersect any scanline of $S$ as any

---

Procedure **ParaCuts**($S$)
**Input** : A set of parallel scanlines $S$.
**Output**: A triangulation of weight at most $4\Delta(G)$ of the polygon defined by the
            scanpoints of the circle model.

> **while** $S$ is not maximal parallel **do**
> > Choose a vertex chord $v$ such that $S \cup \{v\}$ is parallel
> > $S \leftarrow S \cup$ **ScanChord**($S, v$)
>
> Let $\tilde{s}_1$ and $\tilde{s}_2$ be the scanlines delimiting the two *outer polygons*
> $S \leftarrow S \cup$ **TriangOuter**($S, \tilde{s}_1$) $\cup$ **TriangOuter**($S, \tilde{s}_2$)
> **foreach** *inner polygon* $P_{\tilde{t}_1, \tilde{t}_2}$ **do**
> > $S \leftarrow S \cup$ **TriangInner**($S, \tilde{t}_1, \tilde{t}_2$)
>
> **return** $S$

Figure 6.6: Procedure **ParaCuts** computing a triangulation of weight at most $4\Delta(G)$ of
the polygon defined by the scanpoints of the circle model

---

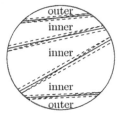

Figure 6.7: Illustration of **ParaCuts**($S$)

scanline intersecting a scanline of $X$ intersects also $v$.

In the procedure **ParaCuts**, the notions of inner and outer polygons are used with respect
to $S$ (see Figure 6.7). In the while-loop, the chosen vertex chord $v$ does not cross a scanline
of $S$ since $S \cup \{v\}$ is required to be parallel. Thus, when the procedure **ScanChord** is called,
its conditions are satisfied. After the while-loop, $S$ is maximal parallel. Every vertex chord
intersecting an outer polygon crosses therefore the scanline delimiting this outer polygon, and
there is no vertex chord between two scanlines delimiting an inner polygon, which are necessary
conditions for **TriangOuter** and **TriangInner**. Moreover, at most $\Delta$ vertex chords cross each
of the delimiting scanlines and no scanline of $S$ intersects the inner and outer polygons.

In the procedure **TriangOuter**, at most $2\Delta$ vertex chords intersect the outer polygon $\mathcal{P}_{\tilde{s}}$.
So, any triangulation of $\mathcal{P}_{\tilde{s}}$ produces triangles with weight at most $2\Delta$. As the procedure
produces a triangulation of $\mathcal{P}_{\tilde{s}}$, it is valid.

Consider the input of the procedure **TriangInner**. There are at most $3\Delta$ vertex chords
inside the quadrilateral $\tilde{a}_1, \tilde{b}_1, \tilde{b}_2, \tilde{a}_2$ since there is no vertex chord crossing both the lines $\tilde{a}_1, \tilde{a}_2$
and $\tilde{b}_1, \tilde{b}_2$ (there is no vertex chord between $\tilde{s}_1$ and $\tilde{s}_2$). As fewer vertex chords cross $\tilde{a}_1, \tilde{a}_2$ than
$\tilde{b}_1, \tilde{b}_2$, at most $3\Delta/2$ vertex chords cross the new scanline $\tilde{t} = \langle \tilde{a}_1, \tilde{a}_2 \rangle$. So, when **OuterPara-**

Procedure **TriangOuter**$(S, \tilde{s} = \langle \tilde{a}, \tilde{b} \rangle)$
**Input**   : A set of scanlines $S$ and a scanline $\tilde{s} \in S$ satisfying the conditions:
             (i) every vertex chord intersecting $\mathcal{P}_{\tilde{s}}$ crosses $\tilde{s}$,
             (ii) at most $2\Delta$ vertex chords cross $\tilde{s}$, and
             (iii) no scanline of $S$ intersects $\mathcal{P}_{\tilde{s}}$.
**Output**: A set of scanlines triangulating the outer polygon $\mathcal{P}_{\tilde{s}}$.

$X \leftarrow \emptyset$
**foreach** scanpoint $\tilde{p}_i \in \mathcal{P}_{\tilde{s}} \setminus \{\tilde{a}, \tilde{b}\}$ **do**
$\quad \lfloor \ X \leftarrow X \cup \{\langle \tilde{a}, \tilde{p}_i \rangle\}$
**return** $X$

Figure 6.8: Procedure **TriangOuter** computing a set of scanlines triangulating an outer polygon where every vertex chord in this polygon crosses its delimiting scanline

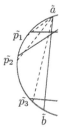

Figure 6.9: Illustration of **TriangOuter**$(S, \tilde{s} = \langle \tilde{a}, \tilde{b} \rangle)$

**Cuts**$(S \cup \{\tilde{t}\}, \tilde{t})$ is called, the condition that $\tilde{t}$ intersects at most $2\Delta$ vertex chords is respected. For every end point $e_i$ of a vertex chord $v_i$ that crosses $\tilde{s}_1$, two triangles are created: $\tilde{a}_1, \tilde{d}_{i-1}, \tilde{d}_i$ and $\tilde{d}_i, \tilde{d}_{i-1}, \tilde{d}'_i$.

The following claim is both the bottleneck and the crucial point of our argument.

**Claim 6.14.** *The triangle* $\tilde{a}_1, \tilde{d}_{i-1}, \tilde{d}_i$ *intersects at most* $4\Delta$ *vertex chords.*

*Proof.* Observe that every vertex chord intersecting this triangle and not crossing $\tilde{s}_1$ crosses either $v_i$ or $v_{i-1}$. As at most $2\Delta$ vertex chords cross $\tilde{s}_1$, at most $\Delta$ cross $v_i$ and at most $\Delta$ cross $v_{i-1}$, the weight of this triangle is at most $4\Delta$. $\qquad \square$

Moreover, at most $2\Delta + 1$ vertex chords cross $\tilde{s}''_i$ and at most $2\Delta$ vertex chords cross $\tilde{s}'''_i$. So, the weight of the triangle $\tilde{d}_i, \tilde{d}_{i-1}, \tilde{d}'_i$ is at most $2\Delta + 1$ and when **OuterParaCuts**$(S \cup X, \tilde{s}''_i)$ is called, the condition that the second parameter of the procedure is a scanline that crosses at most $2\Delta$ vertex chords is respected.
After adding the scanlines $\tilde{s}_3$ and $\tilde{s}_4$ we obtain two more triangles: $\tilde{a}_1, \tilde{d}_k, \tilde{b}_2$ and $\tilde{a}_1, \tilde{b}_2, \tilde{a}_2$. The first one intersects at most $4\Delta$ vertex chords: at most $2\Delta$ cross $\tilde{s}_1$, at most $\Delta$ cross $v_k$ and at most $\Delta$ cross $\tilde{s}_2$. At most $3\Delta$ vertex chords intersect the triangle $\tilde{a}_1, \tilde{b}_2, \tilde{a}_2$: at most $2\Delta$ intersect

Procedure **TriangInner**$(S, \tilde{s}_1 = \langle \tilde{a}_1, \tilde{b}_1 \rangle, \tilde{s}_2 = \langle \tilde{a}_2, \tilde{b}_2 \rangle)$
**Input** : A set of scanlines $S$ and two scanlines $\tilde{s}_1, \tilde{s}_2 \in S$ satisfying the conditions:
        (i) there is no vertex chord between $\tilde{s}_1$ and $\tilde{s}_2$,
        (ii) at most $\Delta$ vertex chords cross one of $\tilde{s}_1$ and $\tilde{s}_2$, say $\tilde{s}_2$,
        (iii) at most $2\Delta$ vertex chords cross the other scanline, $\tilde{s}_1$, and
        (iv) no scanline of $S$ intersects the inner polygon $\mathcal{P}_{\tilde{s}_1, \tilde{s}_2}$.
**Output**: A set of scanlines triangulating $\mathcal{P}_{\tilde{s}_1, \tilde{s}_2}$.

Let the end points of $\tilde{s}_1$ and $\tilde{s}_2$ be ordered $\tilde{a}_1, \tilde{b}_1, \tilde{b}_2, \tilde{a}_2$ around the circle. Assume w.l.o.g., that fewer vertex chords cross the line $\tilde{a}_1, \tilde{a}_2$ than the line $\tilde{b}_1, \tilde{b}_2$
Let $\tilde{t} := \langle \tilde{a}_1, \tilde{a}_2 \rangle$
$X \leftarrow \{\tilde{t}\} \cup \textbf{OuterParaCuts}(S \cup \{\tilde{t}\}, \tilde{t})$
Go around the circle from $\tilde{b}_1$ to $\tilde{b}_2$ (without passing through $\tilde{a}_1$ and $\tilde{a}_2$). Denote by $e_1, \ldots, e_k$ the encountered end points of those vertex chords that cross $\tilde{s}_1$
**foreach** $e_i, i = 1$ to $k$ **do**
    Let $\tilde{s}'_i := \langle \tilde{a}_1, \tilde{d}_i \rangle$ with $\tilde{d}_i$ being the scanpoint following $e_i$
    Let $\tilde{s}''_i := \langle \tilde{d}_i, \tilde{d}_{i-1} \rangle$ with $\tilde{d}_0 = \tilde{b}_1$
    Let $\tilde{s}'''_i := \langle \tilde{d}_{i-1}, \tilde{d}'_i \rangle$ with $\tilde{d}'_i$ being the scanpoint preceding $\tilde{d}_i$
    $X \leftarrow X \cup \{\tilde{s}'_i, \tilde{s}''_i, \tilde{s}'''_i\}$
    $X \leftarrow X \cup \textbf{OuterParaCuts}(S \cup X, \tilde{s}'''_i)$
Let $\tilde{s}_3 := \langle \tilde{d}_k, \tilde{b}_2 \rangle$ and $\tilde{s}_4 := \langle \tilde{b}_2, \tilde{a}_1 \rangle$
$X \leftarrow X \cup \{\tilde{s}_3, \tilde{s}_4\}$
$X \leftarrow X \cup \textbf{OuterParaCuts}(S \cup X, \tilde{s}_3)$
**return** $X$

Figure 6.10: Procedure **TriangInner** computing a set of scanlines triangulating an inner polygon

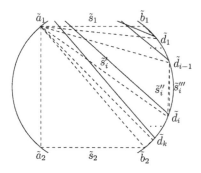

Figure 6.11: Illustration of **TriangInner**$(S, \tilde{s}_1 = \langle \tilde{a}_1, \tilde{b}_1 \rangle, \tilde{s}_2 = \langle \tilde{a}_2, \tilde{b}_2 \rangle)$

$\tilde{s}_1$ and at most $\Delta$ intersect $\tilde{s}_2$. Moreover at most $2\Delta$ vertex chords cross $\tilde{s}_3$. So, the conditions of **OuterParaCuts**$(S \cup X, \tilde{s}_3)$ are respected.

The procedure **OuterParaCuts** is similar to **ParaCuts** on the outer polygon delimited

Procedure **OuterParaCuts**$(S, \tilde{s} = \langle \tilde{a}, \tilde{b} \rangle)$
**Input**   : A set of scanlines $S$ and a scanline $\tilde{s} \in S$ such that
          (i) at most $2\Delta$ vertex chords cross $\tilde{s}$, and
          (ii) no scanline of $S$ intersects $\mathcal{P}_{\tilde{s}}$.
**Output**: A set of scanlines triangulating the outer polygon $\mathcal{P}_{\tilde{s}}$.

  $X \leftarrow \{\tilde{s}\}$
  **while** $X$ is not a maximal parallel in $\mathcal{P}_{\tilde{s}}$ **do**
     Choose a chord $v \in \mathcal{P}_{\tilde{s}}$ such that $X \cup \{v\}$ is parallel
     $X \leftarrow X \cup$ **ScanChord**$(X, v)$

  Let $\tilde{t}$ be the scanline delimiting the recently obtained outer polygon with respect to $X$
  that is a subpolygon of $\mathcal{P}_{\tilde{s}}$
  $X \leftarrow X \cup$ **TriangOuter**$(X, \tilde{t})$
  **foreach** inner polygon $P_{\tilde{t}_1, \tilde{t}_2}$ in $\mathcal{P}_{\tilde{s}}$ **do**
     $X \leftarrow X \cup$ **TriangInner**$(X, \tilde{t}_1, \tilde{t}_2)$
  **return** $X$

Figure 6.12: Procedure **OuterParaCuts** computing a set of scanlines triangulating an outer polygon where not necessarily every vertex chord in this polygon crosses the delimiting scanline of the outer polygon

by $\tilde{s}$. A new set of scanlines $X \leftarrow \{\tilde{s}\}$ is created and is made maximal parallel by calling **ScanChord**. If $\{\tilde{s}\}$ is already maximal parallel, then **TriangOuter**$(X, \tilde{s})$ is called and the conditions of that procedure are respected. If other scanlines had to be added to $X$ to make it maximal parallel, the procedure **TriangOuter**$(X, \tilde{t})$ is called for the outer polygon where $\tilde{t}$ is a scanline of $X$ intersecting at most $\Delta$ vertex chords. Moreover, the procedure **TriangInner**$(X, \tilde{t}_1, \tilde{t}_2)$ is called for the inner polygons. Every scanline delimiting the inner polygons intersects at most $\Delta$ vertex chords, except $\tilde{s}$ that can intersect up to $2\Delta$ vertex chords. So, we respect the condition for **TriangInner** that one scanline intersects at most $\Delta$ vertex chords and the other one at most $2\Delta$.

We have provided a recursive algorithm to triangulate the polygon $\mathcal{P}$ and have shown that the obtained triangulation does not contain triangles intersecting more than $4\Delta$ vertex chords. Thus the corresponding tree decomposition of $G$ has width at most $4\Delta - 1$. $\qquad\square$

## 6.3   Conclusion

The treewidth bounds we showed in this chapter are interesting on their own, but we will also use them in the next two chapters to derive faster exponential time algorithms for different problems. Concerning the bound on sparse graphs, it would be interesting to study the treewidth of 4-regular graphs or graphs with maximum degree 4 more directly. One possible direction could be to try to obtain better bounds on the bisection width of 4-regular graphs. For circle graphs, it would be interesting to know if, for large maximum degree $\Delta$, there are infinitely many graphs with treewidth $4\Delta$ up to a constant additive factor. In other words, is the bound for circle graphs tight? Other results relating treewidth and the maximum degree of graphs belonging to special graph classes are obtained in [BT97, GKLT09].

# Chapter 7

# Domination on Graph Classes

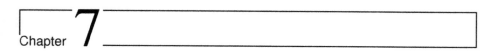

The man who can dominate a London
dinner-table can dominate the world.

*Oscar Wilde*

The MINIMUM DOMINATING SET problem remains $\mathcal{NP}$–hard when restricted to any of the following graph classes: $c$-dense graphs, chordal graphs, 4-chordal graphs, weakly chordal graphs and circle graphs. Developing and using a general approach, for each of these graph classes we present an exponential time algorithm solving MINIMUM DOMINATING SET faster than the best known algorithm for general graphs. Our approach combines a branching algorithm for MINIMUM SET COVER and dynamic programming algorithms for graphs of small treewidth to find a minimum dominating set of graphs belonging to these graph classes. Our algorithms have the following running time: $\mathcal{O}(1.4114^n)$ for chordal graphs, $\mathcal{O}(1.4694^n)$ for weakly chordal graphs, $\mathcal{O}(1.4778^n)$ for 4-chordal graphs, $\mathcal{O}(1.4829^n)$ for circle graphs, and $\mathcal{O}(1.2267^{(1+\sqrt{1-2c})n})$ for $c$-dense graphs.

## 7.1 Related Work

A set $D \subseteq V$ of a graph $G = (V, E)$ is dominating if every vertex of $V \setminus D$ has at least one neighbor in $D$. Given a graph $G = (V, E)$, the MINIMUM DOMINATING SET problem asks to compute a dominating set of minimum cardinality.

Exact exponential time algorithms for the MINIMUM DOMINATING SET problem have not been studied until recently. By now there is a large interest in this particular problem. In 2004 three papers with exact algorithms for MINIMUM DOMINATING SET were published. In [FKW04] Fomin et al. presented an $\mathcal{O}(1.9379^n)$ time algorithm for general graphs and algorithms for split graphs, bipartite graphs and graphs of maximum degree three with running time $\mathcal{O}(1.4143^n)$, $\mathcal{O}(1.7321^n)$, $\mathcal{O}(1.5144^n)$, respectively. Exact algorithms for MINIMUM DOMINATING SET on general graphs have also been given by Randerath and Schiermeyer [RS04] and by Grandoni [Gra06] in 2004. Their running times are $\mathcal{O}(1.8899^n)$ and $\mathcal{O}(1.8026^n)$, respectively.

These algorithms have been significantly improved by Fomin et al. [FGK09b] where the

| graph class | running time |
|---|---|
| $c$-dense graphs | $\mathcal{O}(1.2267^{n(1+\sqrt{1-2c})})$ |
| chordal graphs | $\mathcal{O}(1.4114^n)$ |
| weakly chordal graphs | $\mathcal{O}(1.4694^n)$ |
| 4-chordal graphs | $\mathcal{O}(1.4778^n)$ |
| circle graphs | $\mathcal{O}(1.4829^n)$ |

Table 7.1: Running time of our algorithms for MINIMUM DOMINATING SET on some graph classes

authors obtain faster exact algorithms for MINIMUM DOMINATING SET on general graphs. Their simple branching algorithm is analyzed using a measure–based analysis. Their algorithm has running time $\mathcal{O}(1.5260^n)$ and needs polynomial space. Using memorization one can speed up the running time to $\mathcal{O}(1.5137^n)$ needing exponential space then. Both variants are based on algorithms for the MINIMUM SET COVER problem where the input consists of a universe $\mathcal{U}$ and a collection $\mathcal{S}$ of subsets of $\mathcal{U}$, and the problem requires to find a minimum number of subsets in $\mathcal{S}$ such that their union is equal to $\mathcal{U}$. These algorithms need running time $\mathcal{O}(1.2353^{|\mathcal{U}|+|\mathcal{S}|})$ and polynomial space, or running time $\mathcal{O}(1.2303^{|\mathcal{U}|+|\mathcal{S}|})$ and exponential space.

Van Rooij and Bodlaender [vRB08a] were able to speed up these algorithms and obtained an $\mathcal{O}(1.5134^n)$ time algorithm using polynomial space and an $\mathcal{O}(1.5063^n)$ time algorithm using exponential space. The currently fastest algorithm for MINIMUM DOMINATING SET has been obtained by van Rooij et al. [vRNvD09], building on the previous algorithms and using an inclusion–exclusion branching. This algorithm has running time $\mathcal{O}(1.5048^n)$ and uses exponential space.

Fomin and Høie used a treewidth based approach to establish an algorithm to compute a minimum dominating set for graphs of maximum degree three [FH06] within running time $\mathcal{O}(1.2010^n)$. The best known algorithm for MINIMUM DOMINATING SET on planar graphs has running time $\mathcal{O}(2^{3.99\sqrt{n}})$ [Dor06]. Liedloff [Lie08] constructed an $\mathcal{O}^*(2^{n/2})$ time algorithm to solve MINIMUM DOMINATING SET on bipartite graphs beating the best known algorithm for general graphs.

It is known that MINIMUM DOMINATING SET is $\mathcal{NP}$–hard when restricted to circle graphs [Kei93] and chordal graphs [BJ82], and thus also for weakly chordal and 4-chordal graphs. The $\mathcal{NP}$–hardness of MINIMUM DOMINATING SET for $c$-dense graphs is shown in Section 7.4.

## 7.2    Results

In this chapter we study the MINIMUM DOMINATING SET problem for various graph classes and we obtain algorithms with a running time $\mathcal{O}(\alpha^n)$ better than the best known algorithm solving MINIMUM DOMINATING SET on general graphs. Here the value of $\alpha$ depends on the graph class. We obtain $\alpha < 1.5$ for all classes except for $c$-dense graphs with $c < 0.0155$.

Section 7.3 presents two general frameworks. "Many vertices of high degree" relies heavily on the MINIMUM SET COVER algorithm of van Rooij et al. [vRNvD09]. It is applied to $c$-

dense graphs. Our treewidth based approach uses in fact the "many vertices of high degree" approach for graphs of large treewidth, and otherwise it applies the MINIMUM DOMINATING SET algorithm using a tree decomposition. This approach is applied to chordal, circle, 4-chordal and weakly chordal graphs.

In Section 7.4 we give an $\mathcal{O}(1.2267^{n(1+\sqrt{1-2c})})$ time algorithm for $c$-dense graphs, that is for all graphs with at least $cn^2$ edges, where $c$ is a constant between 0 and 1/2. In Section 7.5 we present exact algorithms solving the MINIMUM DOMINATING SET problem on chordal graphs, weakly chordal graphs, 4-chordal graphs, and circle graphs; see Table 7.1.

The algorithms for circle, 4-chordal and weakly chordal graphs rely on a linear upper bound of the treewidth in terms of the maximum degree. Such bounds are interesting in their own. A related result for graphs of small chordality is provided in [BT97].

## 7.3 General Framework

Our algorithms solve the $\mathcal{NP}$-hard MINIMUM DOMINATING SET problem by exploiting two particular properties of the input graph $G$:

- $G$ has many vertices of high degree:
  $|\{v \in V :\ d(v) \geq t - 2\}| \geq t$ for some (large) positive integer $t$ (see Theorem 7.1);

- there is a constant $c > 0$ such that $\mathbf{tw}(H) \leq c \cdot \Delta(H)$ for all induced subgraphs $H$ of $G$, and there is an algorithm to compute a tree decomposition of $H$ of width at most $c \cdot \Delta(H)$ in polynomial time[1] (see Theorem 7.4).

We describe methods using and combining those properties to establish exponential time algorithms solving MINIMUM DOMINATING SET for a variety of graph classes for which the problem remains $\mathcal{NP}$-hard.

### 7.3.1 Many Vertices of High Degree

The following theorem shows that graphs with sufficiently many vertices of high degree allow to speed up any $\mathcal{O}(\alpha^{2n})$ time algorithm solving MINIMUM DOMINATING SET for general graphs which is based on an algorithm for MINIMUM SET COVER of running time $\mathcal{O}(\alpha^{|\mathcal{U}|+|\mathcal{S}|})$. This is the case for the currently best known algorithm solving MINIMUM DOMINATING SET which is based on an $\mathcal{O}(1.2267^{|\mathcal{U}|+|\mathcal{S}|})$ algorithm for MINIMUM SET COVER [vRB08a], that is $\alpha = 1.2267$.

**Theorem 7.1.** *Suppose there is an $\mathcal{O}(\alpha^{|\mathcal{U}|+|\mathcal{S}|})$ algorithm computing a minimum set cover of any input $(\mathcal{U}, \mathcal{S})$. Let $t(n) : \mathbb{N} \to \mathbb{R}^+$. Then there is an $\mathcal{O}(\alpha^{2n-t(n)})$ time algorithm to solve the MINIMUM DOMINATING SET problem for all input graphs $G$ fulfilling $|\{v \in V :\ d(v) \geq t(n) - 2\}| \geq t(n)$, where $n$ is the number of vertices of $G$.*

---

[1] In fact running time $3^{c \cdot \Delta(H)} \cdot n^{\mathcal{O}(1)}$ suffices.

*Proof.* Let $G = (V, E)$ be a graph fulfilling the conditions of the theorem and let $t := t(n) \geq 0$. Let $T := \{v \in V : d(v) \geq t - 2\}$; thus $|T| \geq t$. Notice that for each minimum dominating set $D$ of $G$ either at least one vertex of $T$ belongs to $D$, or $T \cap D = \emptyset$.

This allows to find a minimum dominating set of $G$ by the following branching in two types of subproblems: "$v \in D$" for each $v \in T$, and "$T \cap D = \emptyset$". Thus we branch into $|T| + 1$ subproblems and for each subproblem we shall apply the $\mathcal{O}(\alpha^{|\mathcal{U}|+|\mathcal{S}|})$ time MINIMUM SET COVER algorithm to solve the subproblems. Recall the transformation given in [FGK09b]: the MINIMUM SET COVER instance corresponding to the instance $G$ for the MINIMUM DOMINATING SET problem has universe $\mathcal{U} = V$ and a collection of sets $\mathcal{S} = \{N[u] : u \in V\}$, and thus $|\mathcal{U}| + |\mathcal{S}| = 2n$. Consequently the running time for a subproblem will be $\mathcal{O}(\alpha^{2n-x})$, where $x$ is the number of elements of the universe plus the number of subsets eliminated from the original MINIMUM SET COVER problem for the graph $G$.

Now let us consider the two types of subproblems. For every vertex $v \in T$, we choose $v$ in the minimum dominating set and we execute the $\mathcal{O}(\alpha^{|\mathcal{U}|+|\mathcal{S}|})$ time MINIMUM SET COVER algorithm on an instance of size at most $2n - (d(v) + 1) - 1 \leq 2n - t$. Indeed, we remove from the universe $\mathcal{U}$ the elements of $N[v]$ and we remove from $\mathcal{S}$ the set corresponding to $v$. When branching into the case "discard $T$" we have an instance of set cover of size at most $2n - |T| = 2n - t$ since for every $v \in T$ we remove from $\mathcal{S}$ the set corresponding to each $v$. □

We would also like to draw the attention to [vRNvD09], where a stronger version of this theorem is proved, needing the set of large–degree vertices only to be half as big.

**Corollary 7.2.** *Let $t(n) : \mathbb{N} \to \mathbb{R}^+$. Then there is an $\mathcal{O}(1.2267^{2n-t(n)})$ time algorithm to solve the* MINIMUM DOMINATING SET *problem for all input graphs $G$ fulfilling $|\{v \in V : d(v) \geq t(n) - 2\}| \geq t(n)$, where $n$ is the number of vertices of $G$.*

### 7.3.2   Treewidth Based Approach

To exploit tree decompositions of small width we rely on the following result of van Rooij et al.

**Theorem 7.3** ([vRBR09]). *There is an algorithm taking as input a graph $G = (V, E)$ and a tree decomposition $T$ of $G$, which computes a minimum dominating set of $G$ in time $\mathcal{O}(3^k k^2 |V|)$, where $k$ is the width of $T$.*

The following theorem shows how to solve the MINIMUM DOMINATING SET problem on a hereditary class of graphs such that a tree decomposition of width at most $c\Delta(G)$ can be obtained for each graph $G$ of the class, where $c$ is a fixed constant. The idea is that such graphs either have many vertices of high degree or their maximum degree is small and thus their treewidth is small. In the first case the algorithm of the previous subsection is used. In the second case the $3^k \cdot n^{\mathcal{O}(1)}$ time algorithm of Theorem 7.3 is used. To balance the running time of the two parts, a parameter $\lambda$ is appropriately chosen.

**Theorem 7.4.** *Suppose there is an $\mathcal{O}^*(\alpha^{|\mathcal{U}|+|\mathcal{S}|})$ algorithm computing a minimum set cover of any input $(\mathcal{U}, \mathcal{S})$. Let $c > 0$ be a constant. Let $\mathcal{G}$ be a hereditary class of graphs such that there is an algorithm that for any input graph $G \in \mathcal{G}$ computes a tree decomposition of width at most*

---

**Algorithm DS-HighDeg-SmallTw(a graph $G = (V, E)$)**
**Input** : A graph $G$ fulfilling the conditions of Theorem 7.4.
**Output**: The domination number $\gamma(G)$ of $G$.

$\lambda \leftarrow \lambda(c, \alpha)$          // the value of $\lambda$ is given in the proof of Theorem 7.4
$X \leftarrow \{u \in V : d(u) \geq \lambda n/c\}$
if $|X| \geq \lambda n/c$ then
  ⌞ use the algorithm of Theorem 7.1 and return the result
else
  ⌞ use the algorithm of Theorem 7.3 and return the result

Figure 7.1: Algorithm for computing the domination number of any graph belonging to a hereditary graph class such that a tree decomposition of width at most $c \cdot \Delta(G)$ of every graph $G$ in this graph class can be computed in polynomial time for some constant $c$

---

| value of $c$ | running time |
|:---:|:---:|
| 1.5 | $\mathcal{O}(1.4629^n)$ |
| 2 | $\mathcal{O}(1.4694^n)$ |
| 2.5 | $\mathcal{O}(1.4741^n)$ |
| 3 | $\mathcal{O}(1.4778^n)$ |
| 4 | $\mathcal{O}(1.4829^n)$ |
| 5 | $\mathcal{O}(1.4865^n)$ |

Table 7.2: Running time of the algorithm in Corollary 7.5 for some values of $c$

$c \cdot \Delta(G)$ *in polynomial time. Then there is an algorithm to solve the* MINIMUM DOMINATING SET *problem for all input graphs of* $\mathcal{G}$ *in time* $\mathcal{O}^*\left(\alpha^{2n \cdot \left(1 - \frac{\log_3 \alpha}{\log_3 \alpha + c + 1}\right)}\right)$.

*Proof.* Let $\lambda = \frac{2 \cdot c \cdot \log_3 \alpha}{\log_3 \alpha + c + 1}$. The algorithm first constructs the vertex set $X$ containing all vertices having degree at least $\lambda n/c$ (see algorithm **DS-HighDeg-SmallTw**).

By definition, for all $v \in X$, $d(v) \geq \lambda n/c$. Thus, if $|X| \geq \lambda n/c$, then we apply the algorithm of Theorem 7.1, and thus a minimum dominating set can be found in time $\mathcal{O}^*(\alpha^{2n - \lambda n/c}) = \mathcal{O}^*\left(\alpha^{2n \cdot (1 - \log_3 \alpha/(\log_3 \alpha + c + 1))}\right)$.

Otherwise $|X| < \lambda n/c$ and $\Delta(G \setminus X) < \lambda n/c$. Note that $G \setminus X$ belongs to the hereditary graph class $\mathcal{G}$ since it is an induced subgraph of $G$. Therefore a tree decomposition of $G \setminus X$ of width at most $c\Delta(G \setminus X) < c\lambda n/c = \lambda n$ can be found in polynomial time. By adding $X$ to every bag of this tree decomposition, one obtains a tree decomposition of $G$ of width at most $\lambda n + \lambda n/c = (c + 1)\lambda n/c$. Now, the algorithm of Theorem 7.3 finds a minimum dominating set in time $\mathcal{O}^*(3^{(c+1)\lambda n/c}) = \mathcal{O}^*\left(\alpha^{2n \cdot (1 - \log_3 \alpha/(\log_3 \alpha + c + 1))}\right)$. ∎

**Corollary 7.5.** *Under the assumptions of Theorem 7.4, there is an algorithm of running time* $\mathcal{O}^*\left(1.2267^{2n \cdot \left(1 - \frac{\log_3 \alpha}{\log_3 \alpha + c + 1}\right)}\right)$ *to solve* MINIMUM DOMINATING SET *for all input graphs of* $\mathcal{G}$.

As we have shown, both methods can be adapted to speed up the algorithms by using any faster MINIMUM SET COVER algorithms established by future work. For different values of $c$, the running time of the algorithm in Corollary 7.5 is displayed in Table 7.2.

In the rest of the chapter we show how the above mentioned general methods can be applied to dense graphs (Section 7.4), chordal graphs, circle graphs, 4-chordal graphs and weakly chordal graphs (Section 7.5).

## 7.4  Dense Graphs

It is known that problems like MAXIMUM INDEPENDENT SET, HAMILTONIAN CIRCUIT and HAMILTONIAN PATH remain $\mathcal{NP}$–hard when restricted to graphs having a large number of edges [Sch95]. In this section we first show that MINIMUM DOMINATING SET also remains $\mathcal{NP}$–hard for $c$-dense graphs. Then we present an exponential time algorithm for the MINIMUM DOMINATING SET problem on this graph class. The algorithm uses the "many vertices of high degree" approach of the previous section.

**Definition 7.6.** A graph $G = (V, E)$ is *c-dense* (or simply dense if there is no ambiguity), if $|E| \geq cn^2$ where $c$ is a constant with $0 < c < 1/2$.

An easy way to show that an $\mathcal{NP}$–hard graph problem remains $\mathcal{NP}$–hard for $c$-dense graphs, for any $c$ with $0 < c < 1/2$, is to construct a graph $G'$ by adding a sufficiently large complete graph as new component to the original graph $G$ such that $G'$ is $c$-dense. This simple reduction can be used to show that various $\mathcal{NP}$–hard graph problems remain $\mathcal{NP}$–hard for $c$-dense graphs. To name a few problems: MAXIMUM INDEPENDENT SET, PARTITION INTO CLIQUES, VERTEX COVER, FEEDBACK VERTEX SET and MINIMUM FILL-IN.

In this way it can be shown that MINIMUM DOMINATING SET is $\mathcal{NP}$–hard for $c$-dense graphs by a polynomial time many–one reduction from the $\mathcal{NP}$–hard problem MINIMUM DOMINATING SET for split graphs.

**Theorem 7.7.** *For any constant $c$ with $0 < c < 1/2$, the problem to decide, whether a $c$-dense graph has a dominating set of size at most $k$ is $\mathcal{NP}$–complete, even when the inputs are restricted to split graphs.*

*Proof.* Let $c$ be any constant with $0 < c < 1/2$. Clearly, the problem to decide whether a graph — and thus also a $c$-dense graph — has a dominating set of size at most $k$ is in $\mathcal{NP}$.

It is shown in [Ber84] that the problem of determining whether a split graph has a dominating set of size at most $k$ is $\mathcal{NP}$-complete. We shall provide a polynomial many–one reduction from MINIMUM DOMINATING SET for split graphs to MINIMUM DOMINATING SET for $c$-dense split graphs.

Let $k$ be an integer and $G = (V = I \cup C, E)$ a split graph where $I$ and $C$ form a partition of the vertices of $G$ such that $I$ is an independent set and $C$ is a clique. First we construct a $c$-dense graph $G' = (V', E')$ with $E' \geq c \cdot |V'|^2$. The graph $G' = (V', E')$ is obtained from the graph $G$ by adding a clique $C'$ of size $\left\lceil (1 + 4c|V| + \sqrt{1 + 8c|V|(1 + |V|)})/(2 - 4c) \right\rceil$ to $G$ and adding all edges with one end point in $C$ and the other in $C'$. This ensures that

$|C'| \cdot (|C'| - 1)/2 \geq c \cdot (|V| + |C'|)^2$. Thus, $G'$ is a split graph with a partition of $V'$ into an independent set $I$ and a clique $C \cup C'$ with at least $c(|V'|)^2$ edges, and hence $G'$ is a $c$-dense split graph.

Now we show that $G$ has a dominating set of size a most $k$ if and only if $G'$ has a dominating set of size at most $k$.

First, assume that $G'$ has a dominating set $D$ with $|D| \leq k$. Since $N_{G'}[x'] \subseteq N_{G'}[x]$ for all $x' \in C'$ and all $x \in C$, we may replace each vertex of $C'$ belonging to $D$ by a vertex of $C$. In this way we obtain a dominating set $D' \subseteq I \cup C$ of $G'$ such that $|D'| \leq k$. Consequently $D'$ is also a dominating set of $G$.

Conversely, assume that $D$ is a dominating set of $G$ of size at most $k$. If $D$ contains at least one vertex of $C$ then $D$ is also a dominating set of $G'$ since each vertex of $C'$ is adjacent to all vertices of $C$. Otherwise, $D$ contains no vertex of $C$ and thus each vertex in $C$ has at least one neighbor in $D \cap I$. In this second case we replace any vertex $s \in D \cap I$ by a neighbor $t \in C$ and obtain $D' = (D \setminus \{s\}) \cup \{t\}$. Then $D'$ is a dominating set of $G$ since $N_G[s] \subseteq N_G[t]$. Furthermore, since $D'$ contains a vertex of $C$ it is also a dominating set of $G'$. Hence $G'$ has in each case a dominating set of size at most $k$.

Thus we obtain that the problem of deciding whether a $c$-dense split graph has a dominating set of size at most $k$ is $\mathcal{NP}$-complete. $\square$

The main idea of our algorithm is to find a large subset of vertices of large degree.

**Lemma 7.8.** *For some fixed* $t, t', 1 \leq t \leq n,\ 1 \leq t' \leq n - 1$, *any graph* $G = (V, E)$ *with* $|E| \geq 1 + \dfrac{(t - 1)(n - 1) + (n - t + 1)(t' - 1)}{2}$ *has a subset* $T \subseteq V$ *such that*

(i) $|T| \geq t$,

(ii) *for every* $v \in T$, $d(v) \geq t'$.

*Proof.* Let $1 \leq t \leq n,\ 1 \leq t' \leq n - 1$, and a graph $G = (V, E)$ such that there is no subset $T$ with the previous properties. Then for any subset $T \subseteq V$ of size at least $t$, there exists a vertex $v \in T$ such that $d(v) < t'$. Then such a graph can only have at most $k := k_1 + k_2$ edges where $k_1 = (t - 1)(n - 1)/2$ (which corresponds to $t - 1$ vertices of degree $n - 1$) and $k_2 = (n - t + 1)(t' - 1)/2$ (which corresponds to $n - (t - 1)$ vertices of degree $t' - 1$). Observe that if one of the $n - (t - 1)$ vertices has a degree greater than $t' - 1$ then the graph has a subset $T$ with the required properties, a contradiction. $\square$

**Lemma 7.9.** *Every* $c$-dense *graph* $G = (V, E)$ *has a set* $T \subseteq V$ *fulfilling*

(i) $|T| \geq \left\lfloor n - \dfrac{\sqrt{9 - 4n + 4n^2 - 8cn^2} - 3}{2} \right\rfloor$,

(ii) *for every* $v \in T$, $d(v) \geq \left\lfloor n - \dfrac{\sqrt{9 - 4n + 4n^2 - 8cn^2} - 3}{2} \right\rfloor - 2$.

*Proof.* We apply Lemma 7.8 with $t' = t - 2$. Since we have a dense graph, $|E| \geq cn^2$. Using inequality $1 + ((t - 1)(n - 1) + (n - t + 1)(t - 3))/2 \leq cn^2$ we obtain that in a dense graph the value of $t$ in Lemma 7.8 is such that $n - \frac{3 - \sqrt{9 - 4n + 4n^2 - 8cn^2}}{2} \geq t$. $\square$

Using the "many vertices of high degree" approach we establish

**Theorem 7.10.** *For any $c$ with $0 < c < 1/2$, there is an $\mathcal{O}\left(1.2267^{(1+\sqrt{1-2c})n}\right)$ time algorithm to solve the* MINIMUM DOMINATING SET *problem on c-dense graphs.*

*Proof.* Combining Theorem 7.1, Corollary 7.2 and Lemma 7.9 we obtain an algorithm for solving the MINIMUM DOMINATING SET problem in time

$$\mathcal{O}\left(1.2267^{2n-\left\lceil n-\frac{\sqrt{9-4n+4n^2-8cn^2}-3}{2}\right\rceil}\right) = \mathcal{O}\left(1.2267^{n+\frac{\sqrt{9-4n+4n^2-8cn^2}-3}{2}}\right)$$
$$= \mathcal{O}\left(1.2267^{n+\frac{\sqrt{9+4n^2(1-2c)}}{2}}\right)$$
$$= \mathcal{O}\left(1.2267^{n+\frac{3+2n\sqrt{1-2c}}{2}}\right)$$
$$= \mathcal{O}\left(1.2267^{n(1+\sqrt{1-2c})} \cdot 1.2267^{\frac{3}{2}}\right)$$
$$= \mathcal{O}\left(1.2267^{n(1+\sqrt{1-2c})}\right).$$

□

## 7.5 Other Graph Classes

In this section, we present exponential time algorithms for the MINIMUM DOMINATING SET problem on chordal graphs, circle graphs, 4-chordal graphs and weakly chordal graphs in a treewidth based approach.

A graph is *chordal* if it has no chordless cycle of length greater than three. Chordal graphs is a well–known graph class with its own chapter in Golumbic's monograph [Gol80]. Split graphs, strongly chordal graphs and undirected path graphs are well–studied subclasses of chordal graphs.

We shall use the clique tree representation of chordal graphs that we view as a tree decomposition of the graph. A tree $T$ is as *clique tree* of a chordal graph $G = (V, E)$ if there is a bijection between the maximal cliques of $G$ and the nodes of $T$ such that for each vertex $v \in V$ the cliques containing $v$ induce a subtree of $T$. It is well–known that $\mathbf{tw}(G) \geq \omega(G) - 1$ for all graphs. Furthermore the clique tree of a chordal graph $G$ is an optimal tree decomposition of $G$, that is its width is precisely $\omega(G) - 1$.

We use the following Corollary of Theorem 7.1.

**Corollary 7.11.** *There is an algorithm taking as input a graph $G$ and a clique $C$ of $G$ and solving the* MINIMUM DOMINATING SET *problem in time* $\mathcal{O}(1.2267^{2n-|C|})$.

*Proof.* Use Corollary 7.2 and note that every vertex in $C$ has degree at least $|C| - 1$.   □

Our algorithm on chordal graphs works as follows: If the graph has large treewidth then it necessarily has a large clique and we apply Corollary 7.11. Otherwise the graph has small treewidth and we use Theorem 7.3.

**Theorem 7.12.** *There is an $\mathcal{O}(1.4114^n)$ time algorithm to solve the* MINIMUM DOMINATING SET *problem on chordal graphs.*

*Proof.* If $\mathbf{tw}(G) < 0.3136n$, by Theorem 7.3, MINIMUM DOMINATING SET can be solved in time $\mathcal{O}(3^{0.3136n}) = \mathcal{O}(1.4114^n)$. Otherwise, $\mathbf{tw}(G) \geq 0.3136n$ and using Corollary 7.11 we obtain an $\mathcal{O}(1.2267^{2n-0.3136n}) = \mathcal{O}(1.4114^n)$ time algorithm. $\qquad\square$

Recall from Section 6.2 that for any circle graph $G$, $\mathbf{tw}(G) \leq 4\Delta(G)$ (Theorem 6.13). Now we simply apply our treewidth based approach of Section 7.3 to circle graphs. Furthermore the class of circle graphs is hereditary and there is a polynomial time algorithm to compute an optimal tree decomposition of circle graphs (Theorem 6.9). Consequently Theorem 7.4 and Corollary 7.5 can be applied and we obtain:

**Theorem 7.13.** *There is an $\mathcal{O}(1.4829^n)$ algorithm to solve* MINIMUM DOMINATING SET *for circle graphs.*

The *chordality* of a graph is the size of its longest chordless cycle. A graph is *4-chordal* if its chordality is at most 4. Thus 4-chordal graphs are a superclass of chordal graphs. We constructively show in [GKLT09] that, for any 4-chordal graph $G$, its treewidth is at most $3\Delta(G)$.

**Theorem 7.14** ([GKLT09]). *For any 4-chordal graph $G$, $\mathbf{tw}(G) \leq 3\Delta(G)$. Moreover, there is a polynomial time algorithm computing, for any 4-chordal input graph $G$, a tree decomposition of width at most $3\Delta(G)$.*

Combining Theorem 7.14 and Corollary 7.5 one establishes:

**Theorem 7.15.** *There is an $\mathcal{O}(1.4778^n)$ algorithm to solve* MINIMUM DOMINATING SET *for 4-chordal graphs.*

A graph $G$ is *weakly chordal* if both $G$ and its complement are 4-chordal. It is easy to check that chordal graphs are a proper subclass of weakly chordal graphs, which are in turn a proper subclass of 4-chordal graphs. The treewidth of weakly chordal graphs can be computed in polynomial time [BT01].

We show in [GKLT09] that for any weakly chordal graph $G$ its treewidth is at most $2\Delta(G)$.

**Theorem 7.16** ([GKLT09]). *For any weakly chordal graph $G$, $\mathbf{tw}(G) \leq 2\Delta(G)$.*

Combining Theorem 7.16 and Corollary 7.5 one establishes:

**Theorem 7.17.** *There is an $\mathcal{O}(1.4694^n)$ algorithm to solve* MINIMUM DOMINATING SET *for weakly chordal graphs.*

## 7.6    Conclusion

We presented several exponential time algorithms to solve the MINIMUM DOMINATING SET problem on graph classes for which this problem remains $\mathcal{NP}$–hard. All these algorithms are faster than the best known algorithm to solve MINIMUM DOMINATING SET on general graphs. We have also shown that any faster algorithm for the MINIMUM SET COVER problem, that is of running time $\mathcal{O}(\alpha^{|\mathcal{U}|+|\mathcal{S}|})$ with $\alpha < 1.2267$, could immediately be used to speed up all our algorithms. It is also clear, that a faster treewidth based algorithm with running time $c^k \cdot n^{\mathcal{O}(1)}$, $c < 3$ taking as input a graph $G = (V, E)$ and a tree decomposition of $G$ of width at most $k$, can be used to speed up the presented algorithms.

Besides classes of sparse graphs (as for example cubic graphs [FH06]) two other graph classes are of interest: split and bipartite graphs. For split graphs, combining ideas of [FKW04] and [vRNvD09] one easily obtains an $\mathcal{O}(1.2267^n)$ algorithm. In [Lie08], Liedloff uses a preprocessing technique to compute a minimum dominating set in time $2^{n-z} \cdot n^{\mathcal{O}(1)}$ of graphs that have an independent set of size $z$, which implies an $\mathcal{O}^*(2^{n/2}) = \mathcal{O}(1.4143^n)$ time algorithm to solve MINIMUM DOMINATING SET on bipartite graphs (see Subsection 1.2.6).

The "high degree" and the "treewidth based" method of this chapter can most likely be applied to other $\mathcal{NP}$–hard problems for constructing fast exponential time algorithms when restricted to graph classes with the corresponding properties. One example is the MINIMUM INDEPENDENT DOMINATING SET problem (see [GL06]).

It is likely that bounds on the treewidth in terms of the maximum degree for circle graphs, 4–chordal graphs, weakly chordal graphs or other graph classes can be used to construct exponential time algorithms for $\mathcal{NP}$–hard problems on special graph classes in a way similar to our approach for domination.

# Chapter 8

# Enumeration and Pathwidth

> It's better to be prepared for an opportunity
> and not have one than to have an opportunity
> and not be prepared.
>
> *Whitney M. Young, Jr*

This chapter presents a generic algorithmic technique based on the enumeration of independent sets and dynamic programming over a path decomposition of the graph. The approach is based on the following idea: either a graph has nice (from the algorithmic point of view) properties which allow a simple recursive procedure to find the solution fast, or the pathwidth of the graph is small, which in turn can be used to find the solution by dynamic programming. By making use of this technique we obtain algorithms

- running in time $\mathcal{O}(1.7272^n)$ for deciding if a graph is 4-colorable,

- running in time $\mathcal{O}(1.6259^n)$ for counting the number of 3-colorings of a graph, and

- running in time $\mathcal{O}(1.4082^n)$ for finding a minimum maximal matching in a graph.

## 8.1   Considered Problems

The CHROMATIC NUMBER problem is one of the oldest and most intensively studied problems in combinatorics and algorithms. The task is to color the vertices of a graph such that no two adjacent vertices are assigned the same color. The smallest number of colors needed to color a graph $G$ is called the *chromatic number*, $\chi(G)$, of $G$. The corresponding decision version of the coloring problem is $k$-COLORING, where for a given graph $G$ and an integer $k$ we are asked if $\chi(G) \leq k$. The $k$-COLORING problem is one of the classical $\mathcal{NP}$-complete problems [GJ79]. In fact it is known to be $\mathcal{NP}$-complete for every $k \geq 3$. A lot of effort was also put in designing efficient approximation algorithms for the optimization version of the problem, namely, given a $k$-colorable graph to try to color it with as few colors as possible. Unfortunately, it has been shown that if certain reasonable complexity conjectures hold then $k$-COLORING is hard to approximate within $n^{1-\varepsilon}$ for any $\varepsilon > 0$ [FK98, KP06].

The history of exponential time algorithms for graph coloring is rich. Christofides obtained the first non–trivial algorithm computing the chromatic number of a graph on $n$ vertices running in time $\mathcal{O}^*(n!)$ in 1971 [Chr71]. In 1976, Lawler [Law76] devised an algorithm with running time $\mathcal{O}(2.4423^n)$ based on dynamic programming over subsets and enumeration of maximal independent sets. Eppstein [Epp03] reduced the bound to $\mathcal{O}(2.4151^n)$ and Byskov [Bys04a] to $\mathcal{O}(2.4023^n)$. In a breakthrough paper, Björklund et al. [BHK09] devised an $\mathcal{O}^*(2^n)$ algorithm for CHROMATIC NUMBER based on a combination of inclusion–exclusion and dynamic programming.

Apart from the general CHROMATIC NUMBER problem, the $k$-COLORING problem for small values of $k$ like 3 and 4 has also attracted a lot of attention. The fastest known algorithm deciding if a the chromatic number of a graph is at most 3 runs in time $\mathcal{O}(1.3289^n)$ and is due to Beigel and Eppstein [BE05]. For 4-COLORING Byskov [Bys04a] designed the previously fastest algorithm, running in time $\mathcal{O}(1.7504^n)$.

The counting version of the $k$-COLORING problem, #$k$-COLORING, is to count the number of all possible $k$-colorings of a given graph. #$k$-COLORING (and its generalization known as CHROMATIC POLYNOMIAL) are among the oldest counting problems. Björklund et al. [BHK09] have also shown that the chromatic polynomial of a graph can be computed in time $\mathcal{O}^*(2^n)$. For $k = 3$, #$k$-COLORING was also studied in the literature. Angelsmark et al. [AJ03] provide an algorithm for #3-COLORING with running time $\mathcal{O}(1.788^n)$. Fürer and Kasiviswanathan [FK05] show how to solve #3-COLORING with running time $\mathcal{O}(1.770^n)$.

In the MINIMUM MAXIMAL MATCHING problem, one is asked to find a maximal matching of minimum size of a graph. For this problem, several exact algorithms can be found in the literature. Randerath and Schiermeyer [RS04] gave an algorithm of time complexity $\mathcal{O}(1.4422^m)$. Raman et al. [RSS07] improved the running time by giving an algorithm of time complexity $\mathcal{O}(1.4422^n)$. They also gave reductions showing that faster algorithms for MINIMUM MAXIMAL MATCHING automatically lead to improved running times for a number of other problems, like MINIMUM EDGE DOMINATING SET and MATRIX DOMINATION.

## 8.2 Our Results

In this chapter we show a generic technique to obtain exact algorithms for several problems for which it is natural to enumerate independent sets. The technique is based on the following combinatorial property which is proved algorithmically and which is interesting in its own: *Either* a graph $G$ has a nice "algorithmic" property which (very sloppily) means that when we apply branching or a recursive procedure to solve a problem then the branching procedure on subproblems of a smaller size works efficiently, *or* (if branching is not efficient) the pathwidth of the graph is small. This type of technique can be used for a variety of problems where the sizes of the subproblems on which the algorithm is called recursively decrease significantly by branching on vertices of high degrees.

In Section 8.3, this technique is presented, along with a general upper bound on the resulting running time, based on the running times of the subprocedures that are plugged into the algorithm. In Section 8.4 we use this technique to obtain exact algorithms for different coloring problems. We show that #3-COLORING and 4-COLORING can be solved in time $\mathcal{O}(1.6259^n)$ and $\mathcal{O}(1.7272^n)$ respectively. These improve the best known results for these two problems. We also apply the technique to MINIMUM MAXIMAL MATCHING and derive an $\mathcal{O}(1.4082^n)$ algorithm for this problem. In [vRB08b], van Rooij and Bodlaender give a faster $\mathcal{O}(1.3226^n)$ algorithm for MINIMUM MAXIMAL MATCHING.

## 8.3 Framework Combining Enumeration and Pathwidth

Let us assume that we have a graph problem for which

(a) we know how to solve it by enumerating independent sets, or maximal independent sets, of the input graph, and

(b) we also know how to solve the problem using dynamic programming over the path decomposition of the input graph.

For example, to check whether a graph $G$ is 3-colorable, one can enumerate all independent sets $I$ of $G$ and for each independent set $I$ can check whether $G\backslash I$ is bipartite. It is also easy to obtain a $3^\ell \cdot n^{\mathcal{O}(1)}$ algorithm for checking if a graph is 3-colorable if a path decomposition of width $\ell$ is known for $G$ (see Lemma 8.5).

For some instances, approach (a) might be faster and for other instances, the path decomposition algorithm might be preferable. One method to get the best of both algorithms would be to compute a path decomposition of the graph using Lemma 6.5 on page 128, and choose one of the two algorithms based on the width of this path decomposition. Unfortunately, this direct method is not very helpful in obtaining better worst case bounds on the running time of the algorithm as it is difficult to predict the running time of the enumeration algorithm for graphs for which the computed path decomposition has large width.

Here in our technique we start by enumerating (maximal) independent sets and based on the knowledge we gain on the graph by this enumeration step, we prove that either the enumeration algorithm is fast, or the pathwidth of the graph is small. This means that either the input graph has a good algorithmic property, or it has a good graph–theoretic property.

To enumerate (maximal) independent sets of the input graph $G$, we use a very standard approach. Two sets $I$ and $C$ are constructed by a recursive procedure, where $I$ is the set of vertices in the independent set and $C$ the set of vertices not in the independent set. Let $v$ be a vertex of maximum degree in $G \setminus (I \cup C)$, the algorithm makes one recursive call where it adds $v$ to $I$ and all its neighbors to $C$ and another recursive call where it adds $v$ to $C$. This branching into two subproblems decreases the number of vertices in $G \setminus (I \cup C)$ according to the following recurrence

$$T(n) \leq T(n - d(v) - 1) + T(n - 1).$$

From this recurrence, we see that the running time of the algorithm depends on how often it branches on a vertex of highest degree. This algorithmic property is reflected by the size of $C$: frequent branchings on vertices of high degree imply that $|C|$ grows fast (in one branch).

Algorithm **enumISPw**$(G, I, C)$
**Input**   : A graph $G$, an independent set $I$ of $G$ and a set of vertices $C$ such that
$N(I) \subseteq C \subseteq V(G) \setminus I$.
**Output**: An optimal solution which has the problem-dependent properties.

**if** $(\Delta(G \setminus (I \cup C)) \geq$ a$)$ or
   $(\Delta(G \setminus (I \cup C)) =$ a$- 1$ and $|C| > \alpha_{\mathsf{a}-1}|V(G)|)$ or
   $(\Delta(G \setminus (I \cup C)) =$ a$- 2$ and $|C| > \alpha_{\mathsf{a}-2}|V(G)|)$ or
   $\cdots$ or
$(\Delta(G \setminus (I \cup C)) = 3$ and $|C| > \alpha_3|V(G)|)$ **then**
   choose a vertex $v \in V(G) \setminus (I \cup C)$ of maximum degree in $G \setminus (I \cup C)$
   $S_1 \leftarrow$ **enumISPw**$(G, I \cup \{v\}, C \cup N(v))$                               R1
   $S_2 \leftarrow$ **enumISPw**$(G, I, C \cup \{v\})$                                          R2
   **return** combine$(S_1, S_2)$
**else if** $\Delta(G \setminus (I \cup C)) = 2$ and $|C| > \alpha_2|V(G)|$ **then**
   **return** enumIS$(G, I, C)$
**else**
   Stop this algorithm and **run** Pw$(G, I, C)$ instead.

Figure 8.1: Algorithm **enumISPw**$(G, I, C)$ combining the approach of enumerating independent sets and of dynamic programming over a path decomposition of the graph to solve various problems

On the other hand we can exploit a graph–theoretic property if $C$ is small and there are no vertices of high degree in $G \setminus (I \cup C)$. In this case we use Lemma 6.5 to upper bound the pathwidth of $G$. If a path decomposition of $G \setminus (I \cup C)$ of size $\beta_d|V(G) \setminus (I \cup C)|$ can be computed, then a path decomposition of $G$ of size $\beta_d|V(G) \setminus (I \cup C)| + |C|$ can be computed easily. Here $\beta_d$ is a constant strictly less than 1 depending on the maximum degree of the graph. If it turns out that a path decomposition of small width can be computed, the algorithm enumerating (maximal) independent sets is completely stopped without any further backtracking and an algorithm based on this path decomposition is executed on the original input graph.

In the rest of this section, we give a general framework combining

- algorithms based on the enumeration of maximal independent sets, and

- algorithms based on path decompositions of small width,

and discuss the running time of the algorithms based on this framework. This framework is not problem–dependent and it relies on two black boxes that have to be replaced by appropriate procedures to solve a specific problem.

Algorithm **enumISPw** $(G, I, C)$ in Figure 8.1 is invoked with the parameters $(G, \emptyset, \emptyset)$, where $G$ is the input graph, and the algorithms enumIS and Pw are problem–dependent subroutines. The function combine is supposed to take polynomial time and it is also a problem–dependent subroutine. The values for a, $\alpha_\mathsf{a}, \ldots, \alpha_3$, and $\alpha_2$ $(0 = \alpha_\mathsf{a} \leq \alpha_{\mathsf{a}-1} \leq \cdots \leq \alpha_2 < 1)$ are carefully chosen constants to balance the time complexities of enumeration and path decomposition based algorithms and to optimize the overall running time of the combined algorithm.

Algorithm enumIS$(G, I, C)$ is problem–dependent and returns an optimal solution *respecting* the choice for $I$ and $C$, where $I$ is an independent set and $C$ is a set of vertices not belonging to the independent set (set of discarded vertices). The sets $I$ and $C$ are usually completed into a (maximal) independent set and a (minimal) vertex cover for $G$ by enumerating (maximal) independent sets of $G \setminus (I \cup C)$, before the problem–specific treatment is done.

Algorithm Pw$(G, I, C)$ first computes a path decomposition based on $G$, $I$ and $C$ and the maximum degree of $G \setminus (I \cup C)$. It then calls a problem–dependent algorithm based on this path decomposition of $G$.

Let $n$ denote the number of vertices of $G$, $T(n)$ be the running time of Algorithm **enumISPw** on $G$, $T_e(n, i, c)$ be the running time of Algorithm enumIS and $T_p(n, i, c)$ be the running time of Algorithm Pw with parameters $G, I, C$ where $i = |I|$ and $c = |C|$. We also assume that for any graph with $n$ vertices and maximum degree $d$, a path decomposition of width at most $\beta_d n$ can be computed. The following lemma is used by Algorithm Pw to compute a path decomposition of $G$ of small width.

**Lemma 8.1.** *Let $\beta_d$ be a constant such that a path decomposition of width at most $\beta_d |V(H)|$ can be computed in polynomial time for any graph $H$ with maximum degree at most $d$. Then a path decomposition of width at most $\beta_d |V(G) \setminus (I \cup C)| + |C|$ can be computed in polynomial time for a graph $G$ if $I$ is an independent set in $G$, $N(I) \subseteq C \subseteq V(G) \setminus I$ and $\Delta(G \setminus (I \cup C)) \leq d$.*

*Proof.* As $I$ is an independent set in $G$ and $C$ separates $I$ from $G \setminus (I \cup C)$, every vertex in $I$ has degree 0 in $G \setminus C$. Thus, a path decomposition of $G \setminus C$ of size at most $\beta_d |V(G) \setminus (I \cup C)|$ can be computed in polynomial time. Adding $C$ to each bag of this path decomposition gives a path decomposition of width at most $\beta_d |V(G) \setminus (I \cup C)| + |C|$ of $G$. □

Given the conditions under which Pw is executed, the following lemma upper bounds its running time.

**Lemma 8.2.** *Let $t_{\mathbf{pw}} > 1$ be a constant. If the considered problem can be solved for any graph $H$ in time $\mathcal{O}^*((t_{\mathbf{pw}})^\ell)$, given a path decomposition of width $\ell$ of $H$, then*

$$T_p(n, i, c) = \mathcal{O}^* \left( \max_{d \in \{2, 3, \dots, a-1\}} \left( (t_{\mathbf{pw}})^{(\beta_d + (1 - \beta_d)\alpha_d) n} \right) \right).$$

*Proof.* The proof follows from Lemma 8.1 and the conditions on $|C|$ and $\Delta(G \setminus (I \cup C))$ under which Algorithm Pw is executed. □

To estimate the maximum size of the search tree we assume that Algorithm Pw is not executed. We denote $\alpha_{d-1} - \alpha_d$ by $\Delta \alpha_d$. Let $t_n, t_i$ and $t_c$ be constants such that $T_e(n, i, c) = \mathcal{O}^*((t_n)^n (t_i)^i (t_c)^c)$. The next lemma bounds the size of the search tree when the algorithm based on a path decomposition is not used.

**Lemma 8.3.** *If Algorithm Pw is not executed, then*

$$T(n) = \mathcal{O}^* \left( (t_n)^n (t_c)^{\alpha_2 n} \prod_{d=3}^{a} (1 + r(d, t_i))^{\Delta \alpha_d n} \right),$$

*where $r(d, t_i)$ is the positive real root of $(1 + x)^{-(d-1)} \cdot x^{-1} \cdot t_i - 1$.*

*Proof.* Let $T_d(n, i, c)$, for $d \in \{2, 3, \ldots, a - 1\}$, be the running time of Algorithm **enumISPw** when Algorithm **Pw** is not executed, and the input of Algorithm **enumISPw** is a triple $(G, I, C)$ with $|V(G)| = n$, $|I| = i$, and $|C| = c$ such that $G \setminus (I \cup C)$ has maximum degree $d$. Also, let $T_a(n, 0, 0) = T(n)$ if Algorithm **Pw** is not executed. Clearly, $T_2(n, i, c) = T_e(n, i, c)$ as Algorithm **Pw** is executed whenever Algorithm **enumIS** is not executed and $G \setminus (I \cup C)$ has maximum degree at most 2. Let us now express $T_d(n, i, c)$ in terms of $T_{d-1}(\cdot, \cdot, \cdot)$ for $d \in \{3, \ldots, a\}$. Consider a node $b$ in the search tree that corresponds to an instance with maximum degree $d < a$ and whose parent corresponds to an instance with maximum degree more than $d$ (if $d = a$, then $b$ is the root of the search tree). Now, we look at the part of the search tree containing $b$ and all descendants of $b$ associated to instances of maximum degree (at least) $d$. Observe that $|C|$ increases by at most $(\alpha_{d-1} - \alpha_d)n = \Delta\alpha_d n$ in this part of the search tree. In each branch of the type **R1**, $|C|$ increases by at least $d$ and in each branch of the type **R2**, $|C|$ increases by 1. We consider a leaf $\ell$ in this part of the search tree. Let $r \in [0, \Delta\alpha_d n/d]$ be the number of times the algorithm branches according to **R1** on the path from $b$ to $\ell$. Then it branches at most $\Delta\alpha_d n - dr$ times according to **R2** on this path. Also note that the distance between $\ell$ and $d$ in the search tree is at most $\Delta\alpha_d n - (d - 1)r$. We get that

$$T_d(n, i, c) = \mathcal{O}^* \left( \sum_{r=0}^{\Delta\alpha_d n/d} \binom{\Delta\alpha_d n - (d-1)r}{r} T_{d-1}(n, i + r, c + \Delta\alpha_d n) \right).$$

In the general situation the degree $d$ may not change to $d - 1$ but rather jump to something smaller. But the worst case bounds on the size of the search tree are achieved when $d$ decreases progressively as considered above.

To prove the lemma, it is sufficient to expand $T_a(n, 0, 0)$ and to prove that

$$\sum_{r=0}^{\Delta\alpha_d n/d} \binom{\Delta\alpha_d n - (d-1)r}{r} t_i^r \tag{8.1}$$

is at most $(1 + r(d, t_i))^{\Delta\alpha_d n}$. The sum in (8.1) is bounded by $(\Delta\alpha_d n/d)B$ where $B$ is the maximum term in this sum. Let us assume that $B = \binom{\Delta\alpha_d n - (d-1)j}{j}(t_i)^j$ for some $j \in \{0, \ldots, \Delta\alpha_d n/d\}$. Now we use the well known fact that for any $x > 0$ and $0 \le k \le n$,

$$\binom{n}{k} \le \frac{(1 + x)^n}{x^k},$$

and we arrive at

$$B = \binom{\Delta\alpha_d n - (d-1)j}{j}(t_i)^j \le \frac{(1 + r(d, t_i))^{\Delta\alpha_d n - (d-1)j}}{r(d, t_i)^j}(t_i)^j$$

$$= (1 + r(d, t_i))^{\Delta\alpha_d n} \cdot \left( \frac{(1 + r(d, t_i))^{-(d-1)}}{r(d, t_i)} \cdot t_i \right)^j = (1 + r(d, t_i))^{\Delta\alpha_d n}.$$

$\square$

The following theorem combines Lemmata 8.2 and 8.3 to upper bound the overall running time of the algorithms resulting from this framework.

**Theorem 8.4.** *The running time of Algorithm* **enumISPw** *on a graph on n vertices is*

$$T(n) = \mathcal{O}^* \left( (t_n)^n (t_c)^{\alpha_2 n} \prod_{d=3}^{a} (1 + r(d, t_i))^{\Delta \alpha_d n} + \max_{d \in \{2,3,\cdots,a-1\}} \left( (t_{\mathbf{pw}})^{(\beta_d + (1-\beta_d)\alpha_d)n} \right) \right) ,$$

*where $r(d, t_i)$ is the positive real root of $(1 + x)^{-(d-1)} \cdot x^{-1} \cdot t_i - 1$.*

The current best values for $\beta_d, 2 \leq d \leq 6$, are obtained from Lemma 6.5 on page 128.

# 8.4 Applications

In this section we use the framework of the previous section to derive algorithms for #3-Coloring, 4-Coloring and Minimum Maximal Matching.

## 8.4.1 Counting 3-Colorings

We first describe the problem-dependent subroutines we need to use in our Algorithm **enumISPw** to solve #3-Coloring in time $\mathcal{O}(1.6259^n)$.

Algorithm **enumISPw** returns here an integer, $I$ corresponds to the color class $C_1$ and $C$ to the remaining two color classes $C_2$ and $C_3$ of every counted coloring of $G$. Algorithm enumIS with parameters $G, I, C$ enumerates all independent sets of $G \setminus (I \cup C)$ and for each, adds this independent set to $I$, and then checks if $G \setminus I$ is bipartite. If $G \setminus I$ is bipartite, then a counter counting the independent sets is incremented by $2^{cc(G \setminus I)}$, where $cc(\cdot)$ denotes the number of connected components of a graph. This takes time $T_e(n, i, c) = 2^{n-i-c}$. Thus, $t_n = 2, t_i = 1/2$ and $t_c = 1/2$.

The function **combine** corresponds in this case to the addition of two integers. The running time of Algorithm Pw is based on the following lemma.

**Lemma 8.5.** *Given a graph $G$ with a path decomposition of $G$ of width $\ell$, #k-Coloring can be solved in time $k^{\ell} n^{\mathcal{O}(1)}$.*

Now we use Theorem 8.4 and Lemma 6.5 to evaluate the overall complexity of our #3-Coloring algorithm.

**Theorem 8.6.** *The #3-Coloring problem can be solved in time $\mathcal{O}(1.6259^n)$.*

*Proof.* We use Theorem 8.4 and Lemma 6.5 with $a = 6, \alpha_2 = 0.4424, \alpha_3 = 0.3308, \alpha_4 = 0.1636$ and $\alpha_5 = 0.0160$. The pathwidth part of the algorithm takes time

$$\mathcal{O}^* \left( \max \left( 3^{\alpha_2 n}, 3^{(1+5\alpha_3)n/6}, 3^{(1+2\alpha_4)n/3}, 3^{13/30+17 \cdot \alpha_5/30} \right) \right) = \mathcal{O}(1.62585^n) .$$

```
> for d from 3 to 6 do
>    t[d] := 1+max(fsolve((1+r)^(-(d-1))*r^(-1)*2^(-1)-1));
> end do;
```

$$t[3] := 1.297156508$$
$$t[4] := 1.253724958$$
$$t[5] := 1.223284957$$
$$t[6] := 1.200511272$$

```
> fsolve({3^a2=3^((1+5*a3)/6),
>         3^a2=3^((1+2*a4)/3),
>         3^a2=3^(13/30+(1-13/30)*a5),
>         3^a2=2*(1/2)^a2*t[3]^(a2-a3)*t[4]^(a3-a4)*
>             t[5]^(a4-a5)*t[6]^a5});
```

```
  {a2 = 0.4423901971, a3 = 0.3308682365,
   a4 = 0.1635852957, a5 = 0.01598270083}
```

Figure 8.2: Maple code for obtaining optimal values for the constants $\alpha_2$ to $\alpha_5$ of Algorithm **enumISPw** for solving the #3-COLORING problem

The branching part of the algorithm takes time

$$\mathcal{O}^* \left( 2^n \cdot (1/2)^{\alpha_2 n} \cdot 1.2972^{(\alpha_2 - \alpha_3)n} \cdot 1.2538^{(\alpha_3 - \alpha_4)n} \cdot 1.2233^{(\alpha_4 - \alpha_5)n} \cdot 1.2006^{\alpha_5 n} \right) = \mathcal{O}(1.62585^n) \ .$$

□

The constants in the proof of Theorem 8.6 are easily obtained via the Maple program in Figure 8.2. Note that it would not help to set a = 7, as $1.6259 < 3^{23/45} \simeq 1.7534$.

## 8.4.2   4-Coloring

A well known technique [Law76] to check if a graph is $k$-colorable is to check for all maximal independent sets $I$ of size at least $\lceil n/k \rceil$ whether $G \setminus I$ is $(k-1)$-colorable. In the analysis, we use the following theorem to bound the number of maximal independent sets of a given size.

**Theorem 8.7** ([Bys04a]). *The maximum number of maximal independent sets of size at most $k$ in any graph on $n$ vertices for $k \leq n/3$ is*

$$N[n,k] := \lfloor n/k \rfloor^{(\lfloor n/k \rfloor + 1)k - n} (\lfloor n/k \rfloor + 1)^{n - \lfloor n/k \rfloor k}.$$

*Moreover, all such sets can be enumerated in time $\mathcal{O}^*(N[n,k])$.*

We also need the currently fastest algorithm deciding 3-COLORING.

**Theorem 8.8** ([BE05]). *3-COLORING can be solved in time $\mathcal{O}(1.3289^n)$.*

In Algorithm **enumISPw**, which returns here a boolean, $I$ corresponds to the color class $C_1$ and $C$ to the remaining three color classes $C_2, C_3$ and $C_4$. Algorithm enumIS with parameters $G, I, C$ enumerates all maximal independent sets of $G \setminus (I \cup C)$ of size at least $\lceil n/4 \rceil - |I|$ and for each, adds this independent set to $I$, then checks if $G \setminus I$ is 3-colorable using Theorem 8.8. If yes, then $G$ is 4-colorable. This takes time

$$T_e(n, i, c) = \sum_{\ell = \lceil n/4 \rceil - i}^{n-i-c} 3^{4\ell - n + c + i} 4^{n - c - i - 3\ell} 1.3289^{n-i-\ell}.$$

As $\sum_{\ell=0}^{\lfloor 3n/4 \rfloor - c} 3^{4\ell} 4^{-3\ell} 1.3289^{-\ell}$ is upper bounded by a constant, $t_n = 4^{1/4} 1.3289^{3/4}$, $t_i = 4^2/3^3$ and $t_c = 3/4$.

**Theorem 8.9.** *The* 4-COLORING *problem can be solved in time* $\mathcal{O}(1.7272^n)$.

*Proof.* We use Theorem 8.4 and Lemma 6.5 with $\mathsf{a} = 5, \alpha_2 = 0.39418, \alpha_3 = 0.27302$ and $\alpha_4 = 0.09127$, and the pathwidth algorithm of Lemma 8.5.                     $\square$

### 8.4.3  Minimum Maximal Matching

Given a graph $G = (V, E)$, any set of pairwise disjoint edges is called a *matching* of $G$. A matching $M$ is maximal if there is no matching $M'$ such that $M \subset M'$. The problem of finding a maximum matching is well studied in algorithms and combinatorial optimization. One can find a matching of maximum size in polynomial time but there are many versions of matching which are $\mathcal{NP}$-hard. Here, we give an exact algorithm for one such version [GJ79]. More precisely, the problem we study is:

> MINIMUM MAXIMAL MATCHING: Given a graph $G = (V, E)$ find a *maximal* matching of *minimum* cardinality.

The enumeration phase, Algorithm enumIS, uses the following characterization of a minimum maximal matching.

**Theorem 8.10** ([RSS07]). *Let* $G = (V, E)$ *be a graph and* $M$ *be a minimum maximal matching of* $G$. *Let*

$$V[M] = \{v \mid v \in V \text{ and } v \text{ is an end point of some edge of } M\}$$

*be a subset of all endpoints of* $M$. *Let* $S \subseteq V[M]$ *be a vertex cover of* $G$. *Let* $M'$ *be a maximum matching in* $G[S]$ *and* $M''$ *be a maximum matching in* $G - V[M']$, *where* $V[M']$ *is the set of the endpoints of edges of* $M'$. *Then* $X = M' \cup M''$ *is a minimum maximal matching of* $G$.

Note that in Theorem 8.10, $S$ does not need to be a *minimal* vertex cover. Therefore Algorithm enumIS enumerates all minimal vertex covers of $G \setminus (I \cup C)$. For every minimal vertex cover $Q$ of $G \setminus (I \cup C)$, $S = C \cup Q$ is a vertex cover of $G$ and the characterization of Theorem 8.10 is used to find a minimum maximal matching of $G$. The running time of this step is within a polynomial factor of the enumeration of all minimal vertex covers of $G \setminus (I \cup C)$, that is $\mathcal{O}(3^{(n-i-c)/3})$ by Theorem 3.5 on page 52 and the following theorem by Johnson et al. [JYP88].

**Theorem 8.11** ([JYP88]). *All maximal independent sets of a graph can be enumerated with polynomial delay.*

For the path decomposition based algorithm, we give a lemma for finding a minimum maximal matching on graphs of bounded pathwidth. The proof of the lemma is based on standard dynamic programming on graphs of bounded pathwidth.

**Lemma 8.12.** *There exists an algorithm to compute a minimum maximal matching of a graph $G$ in time $3^{\ell}n^{\mathcal{O}(1)}$ when a path decomposition of $G$ of width at most $\ell$ is given.*

Plugging all this into our framework, we obtain the following theorem.

**Theorem 8.13.** *A minimum maximal matching of a graph on $n$ vertices can be found in time $\mathcal{O}(1.4082^n)$.*

*Proof.* We use Theorem 8.4 and Lemma 6.5 with $\mathsf{a} = 4, \alpha_2 = 0.31154$ and $\alpha_3 = 0.17385$, the pathwidth algorithm of Lemma 8.12, and the vertex cover enumeration described earlier. □

## 8.5   Conclusion

In this chapter, we combined the enumeration of independent sets and pathwidth based algorithms. The framework we presented is very general in that it suffices to plug in the appropriate subroutines to solve a problem, and the running time analysis is almost immediate via the high–level running time analysis, summarized in Theorem 8.4, that we have provided for the framework. Instead of using pathwidth algorithms as subroutines, one could also combine the enumeration of independent sets with other algorithms that are fast when given a graph $G = (V, E)$ and a small set $C \subseteq V$ such that the maximum degree of $G \setminus C$ is small. Replacing the pathwidth based algorithms by other fast algorithms would be especially interesting if we require only polynomial space usage.

Chapter **9**

# Iterative Compression and Exact Algorithms

> And now for something completely different.
>
> *Monty Python*

Iterative Compression has recently led to a number of breakthroughs in parameterized complexity. The main purpose of this chapter is to show that iterative compression can also be used in the design of exact exponential time algorithms. We exemplify our findings with algorithms for the MAXIMUM INDEPENDENT SET problem, MINIMUM $d$-HITTING SET, #MINIMUM $d$-HITTING SET (a counting version of MINIMUM $d$-HITTING SET) and the MAXIMUM INDUCED CLUSTER SUBGRAPH problem. The algorithms for MINIMUM $d$-HITTING SET with $k \geq 4$, #MINIMUM $d$-HITTING SET with $k \geq 3$ and MAXIMUM INDUCED CLUSTER SUBGRAPH are the currently fastest known algorithms for these problems in the literature.

## 9.1 Background

Iterative Compression is a tool that has recently been used successfully in solving a number of problems in the area of Parameterized Complexity. This technique was first introduced by Reed et al. [RSV04] to solve the ODD CYCLE TRANSVERSAL problem, where one is interested in finding a set of at most $k$ vertices whose deletion makes the graph bipartite. Iterative compression was used in obtaining faster $\mathcal{FPT}$ algorithms for FEEDBACK VERTEX SET, EDGE BIPARTIZATION and CLUSTER VERTEX DELETION on undirected graphs [GGH+06, DFL+07, CFL+08, HKMN08]. This technique has also led to an $\mathcal{FPT}$ algorithm for the DIRECTED FEEDBACK VERTEX SET problem [CLL+08], one of the longest open problems in the area of parameterized complexity.

Typically iterative compression algorithms are designed for parameterized minimization problems with a parameter, say $k$. Such algorithms proceed by iterating the so called *compression step*: given a solution of size $k + 1$, either compress it to a solution of size $k$ or prove that there is no solution of size $k$. To obtain an $\mathcal{FPT}$ algorithm, one has to solve the compression step in time $f(k)n^{O(1)}$, where $f$ is an arbitrary computable function, $k$ is the parameter and $n$ is the length of the input. Technically speaking, almost all iterative compression based $\mathcal{FPT}$

algorithms with parameter $k$ have $f(k) \geq 2^{k+1}$, as they all branch on all partitions $(A, D)$ of a $k+1$ sized solution $S$ and look for a solution of size $k$ with the restriction that it should contain all elements of $A$ and none of $D$.

Given the success of iterative compression in designing $\mathcal{FPT}$ algorithms, it is natural and tempting to study its applicability in designing exact exponential time algorithms solving computationally hard problems.

One simple way to obtain an exact exponential time algorithm from an $\mathcal{FPT}$ algorithm is to use the latter for all possible values of the parameter $k$. In many cases this does not lead to faster exact algorithms. Assuming that the largest (reasonable) value of the parameter $k$ is at least $n$, using a typical iterative compression based $\mathcal{FPT}$ algorithm does not seem to be of much use for constructing an exact exponential time algorithm because we would end up with an algorithm for the compression step having a factor of $2^n$ in its worst-case running time; and hence the established algorithm would not be better than trivial enumeration.

The main advantage of iterative compression is that it provides combinatorial algorithms based on problem structures. While the improvement in the running time compared to (complicated) branching algorithms is not so impressive, the simplicity and elegance of the arguments allow them to be used in a basic algorithm course.

We exemplify this approach by the following results:

1. We show how to solve MAXIMUM INDEPENDENT SET for a graph on $n$ vertices in time $\mathcal{O}(1.3196^n)$. While the running time of our iterative compression algorithm is slower than the running times of branching algorithms [FGK09b, Rob86], this simple algorithm serves as an introductory example to more complicated applications of the method.

2. We obtain the fastest known algorithms counting all minimum hitting sets of a family of sets of an $n$-element ground set, when the size of each set is at most $k \geq 3$ (#MINIMUM $d$-HITTING SET). For the decision problem MINIMUM $d$-HITTING SET, our algorithms are the fastest known algorithms for $k \geq 4$. See Table 9.1 on page 169 for the precise running times of these algorithms. For MINIMUM 4-HITTING SET, our $\mathcal{O}(1.8704^n)$ algorithm improves on previously published algorithms for this problem with running times $\mathcal{O}(1.9646^n)$ [Fer06, RSS07] and $\mathcal{O}(1.9799^n)$ [RSS05].

3. We provide an algorithm to solve the MAXIMUM INDUCED CLUSTER SUBGRAPH problem in time $\mathcal{O}^*(\varphi^n)$ where $\varphi = (1 + \sqrt{5})/2 < 1.6181$ is the golden ratio. The only previous algorithm for this problem we are aware of is the use of a complicated branching algorithm of Wahlström [Wah07] for solving 3-HITTING SET (let us note that MAXIMUM INDUCED CLUSTER SUBGRAPH is a special case of 3-HITTING SET, where every subset is a set of vertices inducing a path of length 3), which results in a running time of $\mathcal{O}(1.6278^n)$.

## 9.2   Maximum Independent Set

MAXIMUM INDEPENDENT SET is a very well studied problem in the area of exact exponential time algorithms and many papers have been written on this problem [TT77, Rob86, Jia86,

ST90, Bei99, FGK09b]. It is customary that if we develop a new method then we first apply it to well known problems in the area. Here, as an introductory example, we consider the $\mathcal{NP}$-complete problem MAXIMUM INDEPENDENT SET.

It is well-known that $I$ is an independent set of a graph $G$ if and only if $V \setminus I$ is a vertex cover of $G$, that is every edge of $G$ has at least one end point in $V \setminus I$. Therefore MINIMUM VERTEX COVER is the complement of MAXIMUM INDEPENDENT SET in the sense that $I$ is a maximum independent set of $G$ if and only if $V \setminus I$ is a minimum vertex cover of $G$. This fact implies that when designing exponential time algorithms we may equivalently consider MINIMUM VERTEX COVER. We proceed by defining a compression version of the MINIMUM VERTEX COVER problem.

---

COMP-MVC: Given a graph $G = (V, E)$ with a vertex cover $S \subseteq V$, find a vertex cover of $G$ of size at most $|S| - 1$ if one exists.

---

Note that if we can solve COMP-MVC efficiently then we can solve MINIMUM VERTEX COVER efficiently by repeatedly applying an algorithm for COMP-MVC as follows. Given a graph $G = (V, E)$ on $n$ vertices with $V = \{v_1, v_2, \ldots, v_n\}$, denote the set $\{v_1, v_2, \ldots, v_i\}$ by $V_i$. Let $G_i := G[V_i]$ and let $C_i$ be a minimum vertex cover of $G_i$. We start with $G_1$ and put $C_1 = \emptyset$. Suppose that we already have computed $C_i$ for the graph $G_i$ for some $i \geq 1$. We form an instance of COMP-MVC with input graph $G_{i+1}$ and $S = C_i \cup \{v_{i+1}\}$. In this stage we either *compress* the solution $S$ which means that we find a vertex cover $S'$ of $G_{i+1}$ of size $|S| - 1$ and put $C_{i+1} := S'$, or (if there is no such $S'$) we put $C_{i+1} := S$.

Our algorithm is based on the following lemma.

**Lemma 9.1.** *Let $G_{i+1}$ and $S$ be given as above. If there exists a vertex cover $C_{i+1}$ of $G_{i+1}$ of size $|S| - 1$, then it can be partitioned into two sets $A$ and $B$ such that*

*1. $A \subset S$ is a minimal vertex cover of $G_{i+1}[S]$.*

*2. $B \subseteq (V_{i+1} \setminus A)$ is a minimum vertex cover of the bipartite graph $G_{i+1} \setminus A$.*

*Proof.* Let $C_{i+1}$ be a vertex cover of $G_{i+1}$ of size $|S| - 1$. Its complement $V_{i+1} \setminus C_{i+1}$ is an independent set. We define $A' = C_{i+1} \cap S$ and $B' = C_{i+1} \setminus A'$. Then $A'$ is a vertex cover of $G_{i+1}[S]$ and $|A'| \leq |S| - 1$. Let $A \subseteq A'$ be a minimal vertex cover of $G_{i+1}[S]$. We define $B = B' \cup (A' \setminus A)$. Since $A$ is a minimal vertex cover of $G_{i+1}[S]$, we have that $S \setminus A$ is an independent set. This in turn implies that $G_{i+1} \setminus A$ is a bipartite graph with bipartition $(S \setminus A, V_{i+1} \setminus S)$. Finally, since $C_{i+1}$ is a minimum vertex cover of $G_{i+1}$, we conclude that $B$ is a minimum vertex cover of $G_{i+1} \setminus A$. □

Lemma 9.1 implies that the following algorithm solves COMP-MVC correctly.

**Step 1** Enumerate all minimal vertex covers of size at most $|S| - 1$ of $G_{i+1}[S]$ as a possible candidate for $A$.

**Step 2** For each minimal vertex cover $A$ find a minimum vertex cover $B$ of the bipartite graph $G_{i+1} \setminus A$ (via the computation of a maximum matching in this bipartite graph [HK73]).

**Step 3** If the algorithm finds a vertex cover $A \cup B$ of size $|S| - 1$ in this way, set $C_{i+1} = A \cup B$, else set $C_{i+1} = S$.

Steps 2 and 3 of the algorithm can be performed in polynomial time, and the running time of Step 1, which is exponential, dominates the running time of the algorithm. To enumerate all maximal independent sets or equivalently all minimal vertex covers of a graph in Step 1, one can use the polynomial–delay algorithm of Johnson et al. [JYP88]; see Theorem 8.11 on page 160.

For the running time analysis of the algorithm we need the $3^{n/3}$ bound on the number of maximal independent sets or minimal vertex covers due to Moon and Moser (Theorem 3.5) and the bound of Byskov (Theorem 8.7) stating that the maximum number of maximal independent sets of size at most $k$ in any graph on $n$ vertices for $k \leq n/3$ is

$$N[n, k] := \lfloor n/k \rfloor^{(\lfloor n/k \rfloor + 1)k - n} (\lfloor n/k \rfloor + 1)^{n - \lfloor n/k \rfloor k},$$

and that they can all be enumerated in time $\mathcal{O}^*(N[n, k])$.
Since

$$\max \left\{ \max_{0 \leq \alpha \leq 3/4} (3^{\alpha n/3}), \max_{3/4 < \alpha \leq 1} (N[\alpha n, (1 - \alpha)n]) \right\} = \mathcal{O}^*(2^{2n/5}),$$

we have that by Theorems 3.5, 8.11, and 8.7, all minimal vertex covers of $G_{i+1}[S]$ of size at most $|S| - 1$ can be listed in time $\mathcal{O}^*(2^{2n/5}) = \mathcal{O}(1.3196^n)$.

Thus, the overall running time of the algorithm solving COMP-MVC is $\mathcal{O}(1.3196^n)$. Since the rounding of the base of the exponent dominates the polynomial factor of the other steps of the iterative compression, we obtain the following theorem.

**Theorem 9.2.** *The given algorithm for the problems* MAXIMUM INDEPENDENT SET *and* MINIMUM VERTEX COVER, *established by iterative compression, has running time* $\mathcal{O}(1.3196^n)$ *on graphs of $n$ vertices.*

## 9.3   #d-Hitting Set

The MINIMUM HITTING SET problem is a generalization of MINIMUM VERTEX COVER. Here, given a family of sets over a ground set of $n$ elements, the objective is to hit every set of the family with as few elements of the ground set as possible. We study a version of the hitting set problem where every set in the family has at most $d$ elements.

---

MINIMUM $d$-HITTING SET : Given a universe $V$ of $n$ elements and a collection $\mathcal{C}$ of subsets of $V$ of size at most $d$, find a minimum hitting set of $\mathcal{C}$. A *hitting set* of $\mathcal{C}$ is a subset $V' \subseteq V$ such that every subset of $\mathcal{C}$ contains at least one element of $V'$.

---

A counting version of the problem is #MINIMUM $d$-HITTING SET that asks for the number of different minimum hitting sets. We denote an instance of #MINIMUM $d$-HITTING SET by

$(V, \mathcal{C})$. Furthermore we assume that for every $v \in V$, there exists at least one set in $\mathcal{C}$ containing it.

We show how to obtain an algorithm to solve #Minimum $d$-Hitting Set using iterative compression which uses an algorithm for #Minimum $(d-1)$-Hitting Set as a subroutine. First we define the compression version of the #Minimum $d$-Hitting Set problem.

Comp-#$d$-Hitting Set: Given a universe $V$ of $n$ elements, a collection $\mathcal{C}$ of subsets of $V$ of size at most $d$, and a (not necessarily minimum) hitting set $H' \subseteq V$ of $\mathcal{C}$, find a minimum hitting set $\widehat{H}$ of $\mathcal{C}$ and compute the number of all minimum hitting sets of $\mathcal{C}$.

**Lemma 9.3.** *Let $\mathcal{O}^* ((a_{d-1})^n)$ be the running time of an algorithm solving #Minimum $(d-1)$-Hitting Set, where $a_{d-1} > 1$ is a constant. Then Comp-#$d$-Hitting Set can be solved in time*

$$\mathcal{O}^* \left( 2^{|H'|} (a_{d-1})^{|V|-|H'|} \right).$$

*Moreover, if $|H'|$ is greater than $2|V|/3$ and the minimum size of a hitting set in $\mathcal{C}$ is at least $|H'| - 1$, then Comp-#$d$-Hitting Set can be solved in time*

$$\mathcal{O}^* \left( \binom{|H'|}{2|H'| - |V|} (a_{d-1})^{|V|-|H'|} \right).$$

*Proof.* To prove the lemma, we give an algorithm that, for each partition $(N, \bar{N})$ of $H'$, computes a minimum hitting set $H_N$ and the number $h_N$ of minimum hitting sets subject to the constraint that these hitting sets contain all the elements of $N$ and none of the elements of $\bar{N}$.

For every partition $(N, \bar{N})$ of $H'$, we either reject it as invalid or we reduce the instance $(V, \mathcal{C})$ to an instance $(V', \mathcal{C}')$ by applying the following two rules in the given order.

**(H)** If there exists a set $C_i \in \mathcal{C}$ such that $C_i \subseteq \bar{N}$ then we refer to such a partition as *invalid* and reject it ($h_N = 0$).

**(R)** For all sets $C_i$ with $C_i \cap N \neq \emptyset$ set $\mathcal{C} := \mathcal{C} \setminus C_i$. In other words, all sets of $\mathcal{C}$, which are already hit by $N$, are removed.

Rule **(H)** is sound because if a set $C_i \subseteq \mathcal{C}$ is included in $\bar{N}$ then no $\widehat{H} \subseteq V \setminus \bar{N}$ contains an element of $C_i$ and thus there is no hitting set $\widehat{H} \subseteq V \setminus \bar{N}$. If a partition $(N, \bar{N})$ of $H'$ is not invalid then the instance $(V, \mathcal{C})$ can be reduced to the instance $I' = (V', \mathcal{C}')$, where $V' = V \setminus H'$ and $\mathcal{C}' = \{X \cap V' \mid X \in \mathcal{C} \text{ and } X \cap N = \emptyset\}$.

Summarizing, the instance $I'$ is obtained by removing all the elements of $V$ for which it has already been decided if they are part of $H_N$ or not and all the sets that are hit by the elements in $N$. To complete $H_N$, it is sufficient to find a minimum hitting set of $I'$ and to count the number of minimum hitting sets of $I'$. The crucial observation here is that $I'$ is an instance of #Minimum $(d-1)$-Hitting Set. Indeed, $H'$ is a hitting set of $(V, \mathcal{C})$ and by removing it we decrease the size of every set by at least one. Therefore, we can use an algorithm for

#Minimum $(d-1)$-Hitting Set to complete this step. When checking all partitions $(N, \bar{N})$ of $H'$ it is straightforward to keep the accounting information necessary to compute a minimum hitting set $\hat{H}$ and to count all minimum hitting sets.

Thus for every partition $(N, \bar{N})$ of $H'$ the algorithm solving #Minimum $(d-1)$-Hitting Set is called for the instance $I'$. There are $2^{|H'|}$ partitions $(N, \bar{N})$ of the vertex set $H'$. For each such partition, the number of elements of the instance $I'$ is $|V'| = |V \setminus H'| = |V| - |H'|$. Thus, the running time of the algorithm is

$$\mathcal{O}^* \left( 2^{|H'|}(a_{d-1})^{|V|-|H'|} \right).$$

If $|H'| > 2|V|/3$ and the minimum size of a hitting set in $\mathcal{C}$ is at least $|H'| - 1$, then it is not necessary to check all partitions $(N, \bar{N})$ of $H'$ and in this case we can speed up the algorithm as the number of relevant partitions of $H'$ becomes significantly smaller than $2^{|H'|}$. Indeed, since

- $|H'| \geq |\hat{H}| \geq |H'| - 1$, and

- $|\hat{H} \cap (V \setminus H')| \leq |V| - |H'|$,

it is sufficient to consider only those partitions $(N, \bar{N})$ of $H'$ such that

$$|N| \geq |H'| - 1 - (|V| - |H'|) = 2|H'| - |V| - 1.$$

In this case, the running time of the algorithm is

$$\mathcal{O}^* \left( \binom{|H'|}{2|H'| - |V|} (a_{d-1})^{|V|-|H'|} \right).$$

This concludes the proof of the lemma.                                                                $\square$

The following lemma will be useful for the forthcoming running-time analysis.

**Lemma 9.4.** *Let $n$ be a natural number and $a$ be a non-negative constant. The sum of the terms $\binom{j}{2j-n} a^{n-j}$ for $j = 0, 1, \ldots, n$ is upper bounded by*

$$\mathcal{O}^* \left( \left( \frac{1 + \sqrt{1 + 4a_{d-1}}}{2} \right)^n \right).$$

*Proof.* Let $j = n - m$. Then $\binom{j}{2j-n} a^{n-j}$ can be rewritten as $g(n, m) = \binom{n-m}{n-2m} a^m = \binom{n-m}{m} a^m$. By convention, we set $g(n, m) = 0$ whenever $n < m$. Denote by $G(n)$ the sum $\sum_{m=0}^{n} g(n, m)$. By a well-known decomposition of binomial coefficients, $g(n, m) = \left( \binom{n-m-1}{m-1} + \binom{n-m-1}{m} \right) a^m = a \cdot g(n-2, m-1) + g(n-1, m)$. Thus $G(n) \leq G(n-1) + a \cdot G(n-2)$. Standard calculus yields that this recurrence is asymptotically upper bounded by $\mathcal{O}^*(\alpha^n)$ where $\alpha$ is the largest positive root of the polynomial $x^2 - x - a = 0$, that is $\alpha = \frac{1+\sqrt{1+4a}}{2}$.                                $\square$

Now we are ready to use iterative compression to prove the following theorem.

**Theorem 9.5.** *Suppose there exists an algorithm to solve* #MINIMUM $(d-1)$-HITTING SET *in time* $\mathcal{O}^*\left((a_{d-1})^n\right)$, $1 < a_{d-1} \le 2$. *Then* #MINIMUM $d$-HITTING SET *can be solved in time*

$$\mathcal{O}^*\left(\left(\frac{1+\sqrt{1+4a_{d-1}}}{2}\right)^n\right).$$

*Proof.* Let $(V, \mathcal{C})$ be an instance of #MINIMUM $d$-HITTING SET, where $V = \{v_1, v_2, \cdots, v_n\}$. For $i = 1, 2, \ldots, n$, let $V_i := \{v_1, v_2, \ldots, v_i\}$ and $\mathcal{C}_i := \{X \in \mathcal{C} \mid X \subseteq V_i\}$. Then $I_i := (V_i, \mathcal{C}_i)$ constitutes an instance for the $i^{th}$ stage of the iteration. We denote by $H_i$ and $h_i$, a minimum hitting set of an instance $I_i$ and the number of different minimum hitting sets of $I_i$ respectively.

If $\{v_1\} \in \mathcal{C}$, then $H_1 = \{v_1\}$ and $h_1 = 1$; otherwise $H_1 = \emptyset$ and $h_1 = 0$.

Consider the $i^{th}$ stage of the iteration. We have that $|H_{i-1}| \le |H_i| \le |H_{i-1}| + 1$ because at least $|H_{i-1}|$ elements are needed to hit all the sets of $I_i$ except those containing element $v_i$ and $H_{i-1} \cup \{v_i\}$ is a hitting set of $I_i$. Now, use Lemma 9.3 with $H' = H_{i-1} \cup \{v_i\}$ to compute a minimum hitting set of $I_i$. If $|H'| \le 2i/3$, its running time is

$$\mathcal{O}^*\left(\max_{0 \le j \le 2i/3} \left\{2^j (a_{d-1})^{i-j}\right\}\right) = \mathcal{O}^*\left(2^{2i/3}(a_{d-1})^{i/3}\right),$$

as $a_{d-1} \le 2$. If $|H'| > 2i/3$, the running time is

$$\mathcal{O}^*\left(\max_{2i/3 < j \le i} \left\{\binom{j}{2j-i}(a_{d-1})^{i-j}\right\}\right).$$

Since for every fixed $j > 2i/3$, and $i$, $1 \le i \le n$,

$$\binom{j}{2j-i}(a_{d-1})^{i-j} \le \binom{j}{2j-n}(a_{d-1})^{n-j},$$

the worst case running time of the algorithm is

$$\mathcal{O}^*\left(\max\left\{\max_{1 \le i \le n} 2^{2i/3}(a_{d-1})^{i/3}, \max_{2n/3 \le j \le n}\left\{\binom{j}{2j-n}(a_{d-1})^{n-j}\right\}\right\}\right).$$

Finally, $\binom{2n/3}{n/3} = 2^{2n/3}$ up to a polynomial factor, and thus the running time is

$$\mathcal{O}^*\left(\max_{2n/3 \le j \le n}\left\{\binom{j}{2j-n}(a_{d-1})^{n-j}\right\}\right) = \mathcal{O}^*\left(\left(\frac{1+\sqrt{1+4a_{d-1}}}{2}\right)^n\right)$$

by Lemma 9.4. $\qquad\square$

Based on the $\mathcal{O}(1.2377^n)$ algorithm for #MINIMUM 2-HITTING SET [Wah07], we have the following corollary.

**Corollary 9.6.** #MINIMUM 3-HITTING SET *can be solved in time* $\mathcal{O}(1.7198^n)$.

The same approach can be used design an algorithm for the optimization version MINIMUM $d$-HITTING SET, assuming that an algorithm for MINIMUM $(d-1)$-HITTING SET is available.

Based on the $\mathcal{O}(1.6278^n)$ algorithm for MINIMUM 3-HITTING SET [Wah07] this leads to an $\mathcal{O}(1.8704^n)$ time algorithm for solving MINIMUM 4-HITTING SET.

**Corollary 9.7.** MINIMUM 4-HITTING SET *can be solved in time* $\mathcal{O}(1.8704^n)$.

In the following theorem we provide an alternative approach to solve #MINIMUM $d$-HITTING SET. This is a combination of brute force enumeration (for sufficiently large hitting sets) with one application of the compression algorithm of Lemma 9.3. For large values of $a_{d-1}$, more precisely for $a_{d-1} \geq 1.6553$, this approach gives faster algorithms than the one obtained by Theorem 9.5.

**Theorem 9.8.** *Suppose there exists an algorithm with running time* $\mathcal{O}^*((a_{d-1})^n)$, $1 < a_{d-1} \leq 2$, *solving* #MINIMUM $(d-1)$-HITTING SET. *Then* #MINIMUM $d$-HITTING SET *can be solved in time*

$$\min_{0.5 \leq \alpha \leq 1} \max \left\{ \mathcal{O}^* \left( \binom{n}{\alpha n} \right), \mathcal{O}^* \left( 2^{\alpha n}(a_{d-1})^{n-\alpha n} \right) \right\}.$$

*Proof.* First the algorithm tries all subsets of $V$ of size $\lfloor \alpha n \rfloor$ and identifies those that are a hitting set of $I$.

Now there are two cases. In the first case, there is no hitting set of this size. Then the algorithm verifies all sets of larger size whether they are hitting sets of $I$. It is straightforward to keep some accounting information to determine the number of hitting sets of the smallest size found during this enumeration phase. The running time of this phase is

$$\mathcal{O}^* \left( \sum_{i=\lfloor \alpha n \rfloor}^{n} \binom{n}{i} \right) = \mathcal{O}^* \left( \binom{n}{\alpha n} \right).$$

In the second case, there exists a hitting set of size $\lfloor \alpha n \rfloor$. Then count all minimum hitting sets using the compression algorithm of Lemma 9.3 with $H'$ being a hitting set of size $\lfloor \alpha n \rfloor$ found by the enumeration phase. By Lemma 9.3, this phase of the algorithm has running time

$$\mathcal{O}^* \left( 2^{\alpha n}(a_{d-1})^{n-\alpha n} \right).$$

By combining the enumeration and compression phases of the algorithm, we obtain the running time claimed in the theorem. $\qquad\square$

The best running times of algorithms solving #MINIMUM $d$-HITTING SET and MINIMUM $d$-HITTING SET are summarized in Table 9.1. For #MINIMUM $d$-HITTING SET, $k \geq 4$, and MINIMUM $d$-HITTING SET, $k \geq 5$, we use the algorithm of Theorem 9.8. Note that the MINIMUM 2-HITTING SET problem is equivalent to MINIMUM VERTEX COVER and MAXIMUM INDEPENDENT SET.

Finally, it is worth to note that the technique can also be used to solve the parameterized version of $d$-Hitting Set. Namely, we consider the following problem:

---

$(k, d)$-HITTING SET: Given a universe $V$ of $n$ elements, a collection $\mathcal{C}$ of subsets of $V$ of size at most $d$ and an integer $k$, find a hitting set of size at most $k$ of $\mathcal{C}$, if one exists.

---

| $d$ | #MINIMUM $d$-HITTING SET | MINIMUM $d$-HITTING SET |
|---|---|---|
| 2 | $\mathcal{O}(1.2377^n)$ [Wah07] | $\mathcal{O}(1.2108^n)$ [Rob86] |
| 3 | $\mathcal{O}(1.7198^n)$ | $\mathcal{O}(1.6278^n)$ [Wah07] |
| 4 | $\mathcal{O}(1.8997^n)$ | $\mathcal{O}(1.8704^n)$ |
| 5 | $\mathcal{O}(1.9594^n)$ | $\mathcal{O}(1.9489^n)$ |
| 6 | $\mathcal{O}(1.9824^n)$ | $\mathcal{O}(1.9781^n)$ |
| 7 | $\mathcal{O}(1.9920^n)$ | $\mathcal{O}(1.9902^n)$ |

Table 9.1: Running times of the algorithms for #MINIMUM $d$-HITTING SET and MINIMUM $d$-HITTING SET

**Theorem 9.9.** *Suppose there exists an algorithm to solve $(k, d-1)$-HITTING SET in time $(a_{d-1})^k \cdot n^{\mathcal{O}(1)}$, where $a_{d-1} \geq 1$. Then $(k, d)$-HITTING SET can be solved in time $(1 + a_{d-1})^k \cdot n^{\mathcal{O}(1)}$.*

*Proof.* The proof is very similar to the one of Theorem 9.5 except that the size of a solution is now bounded by the parameter $k$ instead of $n$. Given a universe $V = \{v_1, v_2, \ldots, v_n\}$ and a collection $\mathcal{C}$, for $i = 1, 2, \ldots, n$, let $V_i = \{v_1, v_2, \ldots, v_i\}$ and $\mathcal{C}_i = \{X \in \mathcal{C} \mid X \subseteq V_i\}$. Then $I_i = (V_i, \mathcal{C}_i)$ constitutes an instance of $(k, d)$-HITTING SET for the $i^{th}$ stage of the iteration. We denote by $H_i$ a hitting set of size at most $k$, if one exists, of the instance $I_i$.

Clearly, if $\{v_1\} \in \mathcal{C}$, then $H_1 = \{v_1\}$ (assuming that $k \geq 1$); otherwise $H_1 = \emptyset$ and $h_1 = 0$.

Consider now the $i^{th}$ stage of the iteration. The relation $|H_{i-1}| \leq |H_i| \leq |H_{i-1}| + 1 \leq k + 1$ holds since at least $|H_{i-1}|$ elements are needed to hit all the sets of $I_i$ except those containing element $v_i$ and $H_{i-1} \cup \{v_i\}$ is a hitting set of $I_i$.

As we did in Lemma 9.3, for every partition $(N, \bar{N})$ of $H_{i-1} \cup \{v_i\}$ we apply the rules **(H)** and **(R)** (see the proof of Lemma 9.3). Each not rejected partition $(N, \bar{N})$ leads to an instance $I_i' = (V_i', \mathcal{C}_i')$ of $(k - |N|, d-1)$-HITTING SET, where $V_i' = V_i \setminus (H_{i-1} \cup \{v_i\})$ and $\mathcal{C}_i' = \{X \cap V_i' \mid X \in \mathcal{C}_i \text{ and } X \cap N = \emptyset\}$. Thus, by using an algorithm for $(k-|N|, d-1)$-HITTING SET, a solution for $I_i'$ of size at most $k - |N|$ can be found, if one exists. The exponential part of the overall running time is given by the formula $\sum_{i=0}^{k} \binom{k+1}{i}(a_{d-1})^{k-i} \leq 2 \cdot (1 + a_{d-1})^k$. $\qquad \square$

In particular, an immediate consequence of the previous theorem is the following.

**Corollary 9.10.** *Suppose there exists an algorithm to solve $(k, 3)$-HITTING SET in time $a_3^k \cdot n^{\mathcal{O}(1)}$. Then, for any fixed $d \geq 4$, $(k, d)$-HITTING SET can be solved in time $(a_3 + d - 3)^k \cdot n^{\mathcal{O}(1)}$.*

By using a $2.0755^k \cdot n^{\mathcal{O}(1)}$ time algorithm by Wahlström [Wah07] for solving $(k, 3)$-HITTING SET, we obtain the running times in Table 9.2 for $d = 4$ and $d = 5$, which are currently the best known algorithms.

# 9.4   Maximum Induced Cluster Subgraph

Clustering objects according to given similarity or distance values is an important problem in computational biology with diverse applications, for example in defining families of orthologous genes, or in the analysis of microarray experiments [DLL+06, FLRS07, Guo09, RWB+07,

| $d$ | $(k,d)$-Hitting Set |
|---|---|
| 2 | $1.2738^k \cdot n^{\mathcal{O}(1)}$ [CKX06] |
| 3 | $2.0755^k \cdot n^{\mathcal{O}(1)}$ [Wah07] |
| 4 | $3.0755^k \cdot n^{\mathcal{O}(1)}$ |
| 5 | $4.0755^k \cdot n^{\mathcal{O}(1)}$ |
| 6 | $5.0640^k \cdot n^{\mathcal{O}(1)}$ [Fer06] |
| 7 | $6.0439^k \cdot n^{\mathcal{O}(1)}$ [Fer06] |

Table 9.2: Running times of the best known algorithms solving $(k,d)$-Hitting Set for various values of $d$. The algorithms for $4 \le d \le 5$ are based on Corollary 9.10.

HKMN08]. A graph theoretic formulation of the clustering problem is called Cluster Editing. To define this problem we need to introduce the notion of a cluster graph. A graph is called a *cluster graph* if it is a disjoint union of cliques. In the most common parameterized version of Cluster Editing, given an input graph $G = (V, E)$ and a parameter $k$, the question is whether the input graph $G$ can be transformed into a cluster graph by adding or deleting at most $k$ edges. This problem has been extensively studied in the realm of parameterized complexity [DLL+06, FLRS07, Guo09, RWB+07]. In this section, we study a vertex version of Cluster Editing. We study the following optimization version of the problem.

Maximum Induced Cluster Subgraph: Given a graph $G = (V, E)$ on $n$ vertices, find a maximum size subset $C \subseteq V$ such that $G[C]$ is a cluster graph.

Due to the following well–known observation, the Maximum Induced Cluster Subgraph problem is also known as Maximum Induced $P_3$-free Subgraph.

**Observation 9.11.** *A graph is a disjoint union of cliques if and only if it contains no induced subgraph isomorphic to the graph $P_3$.*

Clearly, $C \subseteq V$ induces a cluster graph in $G = (V, E)$ (that is $G[C]$ is a disjoint union of cliques of $G$) if and only if $S := V \setminus C$ hits all induced paths on three vertices of $G$. Thus solving the Maximum Induced Cluster Subgraph problem is equivalent to finding a minimum size set of vertices whose removal produces a maximum induced cluster subgraph of $G$. By Observation 9.11, this reduces to finding a minimum hitting set $S$ of the collection of vertex sets of (induced) $P_3$'s of $G$. Such a hitting set $S$ is called a $P_3$-HS.

As customary when using iterative compression, we first define a compression version of the Maximum Induced Cluster Subgraph problem.

Comp-MICS: Given a graph $G = (V, E)$ on $n$ vertices and a $P_3$-HS $S \subseteq V$, find a $P_3$-HS of $G$ of size at most $|S| - 1$ if one exists.

**Theorem 9.12.** COMP-MICS *can be solved in time* $\mathcal{O}(\varphi^n)$ *where* $\varphi = (1 + \sqrt{5})/2 < 1.6181$ *is the golden ratio.*

*Proof.* For the proof we distinguish two cases based on the size of $S$.

**Case 1:** If $|S| \leq 2n/3$ then the following algorithm which uses matching techniques is applied.

**Step 1** Enumerate all partitions of $(N, \bar{N})$ of $S$.

**Step 2** For each partition, compute a maximum set $C \subseteq V$ such that $G[C]$ is a cluster graph, subject to the constraints that $N \subseteq C$ and $\bar{N} \cap C = \emptyset$, if such a set $C$ exists.

In Step 2, we reduce the problem of finding a maximum sized $C$ to the problem of finding a maximum weight matching in an auxiliary bipartite graph.[1]

If $G[N]$ contains an induced $P_3$ then there is obviously no $C \subseteq V$ inducing a cluster graph that respects the partition $(N, \bar{N})$. We call such a partition *invalid*.

Otherwise, $G[N]$ is a cluster graph, and thus the goal is to find a maximum size subset $C'$ of $\bar{S} := V \setminus S$ such that $G[C' \cup N]$ is a cluster graph. Fortunately, such a set $C'$ can be computed in polynomial time by reducing the problem to finding a maximum weight matching in an auxiliary bipartite graph.

First we describe the construction of the bipartite graph. Consider the graph $G[N \cup \bar{S}]$ and note that $G[N]$ and $G[\bar{S}]$ are cluster graphs. Now the following reduction rule is applied to the graph $G[N \cup \bar{S}]$.

**(R)** Remove every vertex $b \in \bar{S}$ for which $G[N \cup \{b\}]$ contains an induced $P_3$.

Clearly all vertices removed by **(R)** cannot belong to any $C'$ inducing a cluster subgraph of $G$. Let $\hat{S}$ be the subset of vertices of $\bar{S}$ which are not removed by **(R)**. Hence the current graph is $G[N \cup \hat{S}]$. Clearly $G[\hat{S}]$ is a cluster graph since $G[\bar{S}]$ is one. Further, note that no vertex of $\hat{S}$ has neighbors in two different maximal cliques of $G[N]$ and if a vertex of $\hat{S}$ has a neighbor in one maximal clique of $G[N]$ then it is adjacent to each vertex of this maximal clique. Thus, every vertex in $\hat{S}$ has either no neighbor in $N$ or it is adjacent to all the vertices of exactly one maximal clique of $G[N]$.

Now we are ready to define the auxiliary bipartite graph $G' = (A, B, E')$. Let $\{\mathcal{C}_1, \mathcal{C}_2, \ldots, \mathcal{C}_r\}$ be the maximal cliques of the cluster graph $G[N]$. Let $\{\mathcal{C}'_1, \mathcal{C}'_2, \ldots, \mathcal{C}'_s\}$ be the maximal cliques of the cluster graph $G[\hat{S}]$. Let $A := \{a_1, a_2, \ldots, a_r, a'_1, a'_2, \ldots, a'_s\}$ and $B := \{b_1, b_2, \ldots, b_s\}$. Here, for all $i \in \{1, 2, \ldots, r\}$, each maximal clique $\mathcal{C}_i$ of $G[N]$ is represented by $a_i \in A$; and for all $j \in \{1, 2, \ldots, s\}$, each maximal clique $\mathcal{C}'_j$ of $G[\hat{S}]$ is represented by $a'_j \in A$ and by $b_j \in B$.

Now there are two types of edges in $G'$: $a_j b_k \in E'$ if there is a vertex $u \in \mathcal{C}'_k$ such that $u$ has a neighbor in $\mathcal{C}_j$, and $a'_j b_j \in E'$ if there is a vertex $u \in \mathcal{C}'_j$ such that $u$ has no neighbor in $N$. Finally we define the weights for both types of edges in the bipartite graph $G'$. For an edge $a_j b_k \in E'$, its weight $w(a_j b_k)$ is the number of vertices in $\mathcal{C}'_k$ being adjacent to all vertices of the

[1]Hüffner et al. [HKMN08] obtain among others an $\mathcal{FPT}$ algorithm for the vertex weighted version of CLUSTER VERTEX DELETION using iterative compression. In their compression step they use the natural idea of reduction to weighted bipartite matching as well.

maximal clique $C_j$. For an edge $a'_j b_j$, its weight $w(a'_j b_j)$ is the number of vertices in $C'_j$ without any neighbor in $N$.

This transformation is of interest due to the following claim that uses the above notation.

**Claim 9.13.** *The maximum size of a subset $C'$ of $\hat{S}$ such that $G[N \cup C']$ is a cluster subgraph of the graph $G^* = G[N \cup \hat{S}]$ is equal to the maximum total weight of a matching in the bipartite graph $G' = (A, B, E')$.*

*Proof.* We first show that any matching in $G'$ corresponds to a set $Y \subseteq \hat{S}$ that together with $N$ induces a cluster subgraph of $G^*$, that is, $G[N \cup Y]$ is a $P_3$-free graph. To see this, let $M := \{e_1, e_2, \ldots, e_t\}$ be a matching in $G'$. Now if $e_l = a_j b_k$ then $Y_l$ is the set of vertices in $C'_k$ which are adjacent to all vertices of the maximal clique $C_j$. Otherwise, if $e_l = a'_j b_j$ then $Y_l$ is the set of vertices in $C'_j$ which have no neighbor in $N$. Now let us set $Y := \bigcup_{l=1}^t Y_l$. Clearly, $|Y| = \sum_{l=1}^t w(e_l)$. We claim that $G[N \cup Y]$ is a disjoint union of cliques. To the contrary, suppose there exists an induced $P_3$ in $G[N \cup Y]$, say $P = xyz$ is an induced $P_3$ in $G[N \cup Y]$. Then two of the vertices of $P$ are in $Y$ and one in $N$ because of rule **(R)** and the fact that $G[\hat{S}]$ is a cluster graph. First let $x, z \in Y$, $y \in N$, $x \in C'_{t_1}$, $y \in C_{t_2}$, and $z \in C'_{t_3}$. This means selecting edges $a_{t_2} b_{t_1}$ and $a_{t_2} b_{t_3}$ in $M$. Secondly, let $x, y \in Y$ and $z \in N$, and thus $x$ and $y$ belong to the same clique $C'_{t_1}$, and $z \in C_{t_2}$. This means having edges $a_{t_2} b_{t_1}$ and $a'_{t_1} b_{t_1}$ in $M$. In both cases this contradicts $M$ being a matching. Consequently if there is a matching $M'$ in $G^*$ of weight $k$ then there is a set $Y \subseteq \hat{S}$ of size $k$ such that $G[N \cup Y]$ is a cluster graph.

To prove the other direction, let $\{\mathcal{F}_1, \mathcal{F}_2, \ldots, \mathcal{F}_q\}$ be the maximal cliques of the cluster graph $G[C']$, and let $\{\mathcal{F}'_1, \mathcal{F}'_2, \ldots, \mathcal{F}'_p\}$ be the maximal cliques of the cluster graph $G[N \cup C']$. Clearly, each $\mathcal{F}'_j$, $1 \leq j \leq p$, contains at most one of $\{\mathcal{F}_l : 1 \leq l \leq q\}$. Let $\pi(l)$ be the integer such that $\mathcal{F}_l \subseteq \mathcal{F}'_{\pi(l)}$. If $\mathcal{F}_l = \mathcal{F}'_{\pi(l)}$ then set $e_l := a'_l b_l$. Otherwise, if $\mathcal{F}_l \subset \mathcal{F}'_{\pi(l)}$ then set $e_l := a_{\pi(l)} b_l$. Since $\pi$ is injective, $M = \{e_1, e_2, \ldots, e_q\}$ is a matching in $G'$ and the definition of the weights of the edges in $G'$ implies that the total weight of $M$ is $\sum_{l=1}^q w(e_l) = |C'|$. Thus there is a matching of $G'$ of total weight $|C'|$.  $\square$

Note that the construction of the bipartite graph $G'$, including the application of **(R)** and the computation of a maximum weighted matching of $G'$ can be performed in time $\mathcal{O}(n^3)$ [EK72]. Thus, the running time of the algorithm in Case 1 is the time needed to enumerate all subsets of $S$ (whose size is bounded by $2n/3$) and this is $\mathcal{O}^*(2^{2n/3}) = \mathcal{O}(1.5875^n)$.

**Case 2:** If $|S| > 2n/3$ then the algorithm needs to find a $P_3$-HS of $G$ of size $|S| - 1$, or show that none exists.

The algorithm proceeds as in the first case. Note that at most $n - |S|$ vertices of $V \setminus S$ can be added to $N$. Therefore, the algorithm verifies only those partitions $(N, \bar{N})$ of $S$ satisfying $|N| \geq |S| - 1 - (n - |S|) = 2|S| - n - 1$. In this second case, the worst-case running time is

$$\mathcal{O}^* \left( \max_{2/3 < \alpha \leq 1} \left\{ \binom{\alpha n}{(2\alpha - 1)n} \right\} \right) = \mathcal{O}^* \left( \left( \frac{1 + \sqrt{5}}{2} \right)^n \right)$$

by Lemma 9.4.  $\square$

Now we are ready to prove the following theorem using iterative compression.

**Theorem 9.14.** MAXIMUM INDUCED CLUSTER SUBGRAPH *can be solved in time* $\mathcal{O}^*(\varphi^n)$ *where* $\varphi = (1 + \sqrt{5})/2 < 1.6181$ *is the golden ratio.*

*Proof.* Given a graph $G = (V, E)$ with $V = \{v_1, v_2, \ldots, v_n\}$, let $V_i := \{v_1, v_2, \ldots, v_i\}$, $G_i := G[V_i]$ and let $C_i$ be a maximum induced cluster subgraph of $G_i$. Let $S_i := V_i \setminus C_i$.

The algorithm starts with $G_1$, $C_1 = \{v_1\}$ and $S_1 = \emptyset$. At the $i^{th}$ iteration of the algorithm, $1 \le i \le n$, we maintain the invariant that we have at our disposal $C_{i-1}$, a maximum set inducing a cluster subgraph of $G_{i-1}$, and $S_{i-1}$, a minimum $P_3$-HS of $G_{i-1}$. Note that $S_{i-1} \cup \{v_i\}$ is a $P_3$-HS of $G_i$ and that no $P_3$-HS of $G_i$ has size smaller than $|S_{i-1}|$. Now use the algorithm of Lemma 9.12 to solve COMP-MICS on $G_i$ with $S = S_{i-1} \cup \{v_i\}$. Then the worst case running time is attained at the $n^{th}$ stage of the iteration and the running time is $\mathcal{O}^*(\varphi^n)$ where $\varphi = (1 + \sqrt{5})/2$. $\square$

## 9.5 Conclusion

Iterative compression is a technique which is successfully used in the design of $\mathcal{FPT}$ algorithms. In this chapter we show that this technique can also be used to design exact exponential time algorithms. This suggests that it might be used in other areas of algorithms as well. For example, how useful can iterative compression be in the design of approximation algorithms?

Carrying over techniques from the design of $\mathcal{FPT}$ algorithms to the design of exact exponential time algorithms and vice–versa is a natural and tempting idea. A challenging question in this regard is whether measure based analyses can fully be adapted for the analysis of $\mathcal{FPT}$ branching algorithms. First results on this topic are discussed in [Gas09].

A question related to the minimization and counting problems of MINIMUM $d$-HITTING SET is to upper bound the maximum number of minimal $d$-hitting sets. A simple branching algorithm [Gas05] shows that this number can be upper bounded by $\mathcal{O}(c_k^n)$ where $c_k$ is the positive real root of $\sum_{i=1}^{k} x^{-i} - 1$.

**Open Question.** *Upper bound the number of minimal d-hitting sets by $c'_k$ where $c'_k < c_k$ for all or some values of $k \ge 3$.*

# Chapter 10

# Conclusion

> With every new answer unfolded, science has consistently discovered at least three new questions.
>
> *Wernher von Braun*

In this text we designed exponential time algorithms for various hard problems and proved upper bounds on the running time of these algorithms. For some of them we provided lower bounds on their worst-case running time. Besides decision and optimization problems, we also considered counting and enumeration problems. General methods to design exponential time algorithms have been developed by combining existing methods, like branching, enumeration of objects and treewidth based algorithms. Moreover, the technique of iterative compression has been used to design faster exponential time algorithms.

As we demonstrated in Section 3.5, methods to analyze exponential time algorithms can also be used to derive combinatorial bounds on mathematical objects. Another example of this is an algorithmic proof of the famous Moon–Moser Theorem (Theorem 3.5) stating that the number of maximal independent sets in a graph $G = (V, E)$ on $n$ vertices is at most $3^{n/3}$. In a Dagstuhl seminar, Kratsch [Kra07] presented the following outline of an alternative proof for this theorem by an algorithm for enumerating all maximal independent sets. If $V \neq \emptyset$, then choose a vertex $v$ of minimum degree in $G$ and for each $u \in N[v]$, add $u$ to each set produced by a recursive call on the instance $G \setminus N[u]$. To bound the number of leafs in the search tree of the algorithm, the strongest constraint

$$(\forall d : 0 \leq d \leq n - 1) \quad T(n) \geq (d + 1) \cdot T(n - (d + 1))$$

is obtained for branching on vertices of degree $d = 2$ and gives the upper bound of $3^{n/3}$ on the maximum number of maximal independent sets in a graph on $n$ vertices.

One possible topic of further research would be the design of exponential or subexponential time approximation algorithms. For the BANDWIDTH problem, Fürer et al. [FGK09c] present a factor 2 approximation algorithm with running time $\mathcal{O}(1.9797^n)$. In this line, the natural question arises asking which approximation guarantees can be achieved in subexponential time under some reasonable complexity–theoretic assumptions.

**Open Question.** *Find a subexponential factor* 2 *approximation algorithm for the* BANDWIDTH *problem, or prove that none can exist under the Exponential Time Hypothesis.*

The obvious difficulty for branching algorithms that was already mentioned a few times is that the analysis tools that are currently available do usually not provide provably tight bounds on the running times of algorithms. Both the derived upper bounds and the lower bounds on the running times of these algorithms might not be tight. How and if this question will be settled for specific algorithms, or in general, is difficult to predict and is a real challenge.

Lower bounds for the worst–case running time of branching algorithms are usually achieved by exhibiting explicit instances, generated from small graphs, for which the algorithm has high running time. Are there other methods to prove lower bounds on the running time of these algorithms? Here methods similar to the measure–based analysis, but working in a best–case scenario instead of a worst–case scenario, together with constraints saying which local configurations may be selected after which branching, could maybe help narrowing the gap between lower and upper bounds on the worst–case running time of an algorithm.

Another further research direction that has been partly pioneered [Rob01, FK04, Kul05] is automated algorithm design and analysis for branching algorithms. In my opinion, the progress that has been made so far does not yet allow a measure of problem instances that is flexible enough. By fixing a measure, it is easy to make a program run through all possible local configurations and identify the worst ones. By fixing a set of local configurations, it is also easy to make a program compute an optimal measure. The main problem is here that ideally, we do not wish to fix either the measure, nor the set of local configurations. A possibility might however be to fix an initial set of local configurations, compute an optimal measure for them, identify the worst local configurations, propose better strategies for these hard cases, and automatically or with human interaction, try to improve them. A considerable effort would ideally be spent to simplify the case analysis as much as possible and the correctness of the resulting algorithm and its analysis would be automatically established, for example using the B Method [Abr96].

# Glossary

**biclique:** Given a graph $G = (V, E)$ and a vertex set $V' \subseteq V$, $V'$ is an (induced) biclique of $G$ if and only if $G[V']$ is complete bipartite.

**bipartite:** A graph $G = (V, E)$ is bipartite if and only if its vertex set can be partitioned into two independent sets. A partition $(A, B)$ of $V$ into independent sets is called a bipartition of $G$. $G$ is then often denoted by $G = (A, B, E)$.

**bipartite complement:** The bipartite complement of a bipartite graph $G = (A, B, E)$ is a bipartite graph having the vertices of $G$ as its vertex set and the non-edges of $G$ with an endpoint in $A$ and another in $B$ as its edge set.

**clique:** Given a graph $G = (V, E)$, a clique in $G$ is a subset $V' \subseteq V$ of vertices such that $G[V']$ induces a complete graph.

**clique number:** The clique number $w(G)$ of a graph $G$ is the size of the largest clique in $G$.

**closed $k$-neighborhood:** The closed $k$-neighborhood of a vertex $v$ in a graph $G$ is $N_G^k[v] := \bigcup_{i=0..k} N_G^i(v)$.

**closed neighborhood:** The closed neighborhood of a vertex $v$ in a graph $G$ is $N_G[v] := \{u\} \cup N_G(v)$.

**coloring:** Given a graph $G = (V, E)$, a coloring of $V$ is a function from $V$ to a set of colors (integers) such that every two adjacent vertices in $G$ are mapped to a different color. A $k$-coloring is a coloring using exactly $k$ colors.

**complete:** A graph $G$ is complete if and only if there is an edge between each pair of vertices in $G$. A complete graph on $n$ vertices is denoted by $K_n$.

**complete bipartite:** A graph is complete bipartite if and only if it is bipartite with bipartition $(A, B)$ and every vertex of $A$ is adjacent to every vertex of $B$. A complete bipartite graph such that the sets of its bipartition have size $x$ and $y$ is denoted $K_{x,y}$.

**connected:** A graph $G$ is connected if there is a walk between every two vertices of $G$.

**connected component:** Maximal connected subgraph.

**cycle:** 2-regular connected graph. A cycle on $n$ vertices is denoted $C_n$.

**degree:** The degree of a vertex $v$ in a graph $G$ is $d_G(v) := |N_G(v)|$.

**distance:** Given a graph $G$ and two vertices $u, v \in V$, the distance between $u$ and $v$ is the length of the shortest walk minus one between $u$ and $v$, that is the minimum number of edges needed to be traversed to reach $v$ from $u$ and is denoted by $dist_G(u, v)$.

**dominating set:** Given a graph $G = (V, E)$ and a vertex set $V' \subseteq V$, $V'$ is a dominating set of $G$ if and only if every vertex in $V \setminus V'$ has a neighbor in $V'$.

**domination number:** The domination number $\gamma(G)$ of a graph $G$ is the size of the smallest dominating set in $G$.

**dual degree:** The dual degree of a vertex $v$ in a graph $G$ is the sum of the degrees of its neighbors $\sum_{u \in N(v)} d_G(u)$.

**feedback vertex set:** Given a graph $G = (V, E)$ and a vertex set $V' \subseteq V$, $V'$ is a feedback vertex set of $G$ if and only if $G \setminus V'$ is a forest.

**forest:** Acyclic graph.

**graph:** A (simple, undirected) graph $G$ is an ordered pair $(V, E)$ of a set $V$ of vertices and a set $E$ of edges, where $E$ is a set of unordered pairs of distinct vertices. Its vertex set is $V(G) = V$ and its edge set is $E(G) = E$.

**independent set:** Given a graph $G = (V, E)$ and a vertex set $V' \subseteq V$, $V'$ is an independent set of $G$ if and only if $G[V']$ has no edges.

**induced subgraph:** Given a graph $G = (V, E)$ and a vertex set $V' \subseteq V$, the subgraph of $G$ induced on $V'$ is the graph $G[V'] := (V', \{uv \in E : u, v \in V'\})$.

**isomorphism:** An isomorphism of two graphs $G$ and $H$ is a bijection between their vertex sets $f : V(G) \to V(H)$ such that $uv \in E(G)$ if and only if $f(u)f(v) \in E(H)$.

**logical Kronecker delta:** $K_\delta(\cdot)$ returns 1 if its argument is true and 0 otherwise.

**maximum degree:** The maximum degree of a graph $G = (V, E)$ is $\Delta(G) := \max_{v \in V} d_G(v)$.

**minimum degree:** The minimum degree of a graph $G = (V, E)$ is $\delta(G) := \min_{v \in V} d_G(v)$.

**open $k$-neighborhood:** The (open) $k$-neighborhood of a vertex $v$ in a graph $G = (V, E)$ is $N_G^k(v) := \{u \in V : u \text{ is at distance } k \text{ from } v\}$.

**open neighborhood:** The (open) neighborhood of a vertex $v$ in a graph $G = (V, E)$ is
$N_G(v) := \{u \in V : uv \in E\}$.

**path:** Tree with maximum degree 2. A path on $n$ vertices is denoted $P_n$. The word path may
also refer to a walk with no repeated vertices.

**regular:** A graph is $d$-regular if each of its vertices has degree $d$. A graph is regular if it is
$d$-regular for some $d$.

**subgraph:** A graph $H$ is a subgraph of a graph $G$ if $V(H) \subseteq V(G)$ and $E(H) \subseteq E(G)$.

**tree:** Acyclic, connected graph.

**vertex cover:** Given a graph $G = (V, E)$, a vertex set $V' \subseteq V$ is a vertex cover of $G$ if and
only if each edge of $G$ is incident to at least one vertex of $V'$.

**vertex removal:** Given a graph $G = (V, E)$ and a vertex set $V' \subseteq V$, the graph obtained from
removing $V'$ from $G$ is $G \setminus V' := G[V \setminus V']$. If $V' = \{u\}$, we may write $G \setminus u$ instead of
$G \setminus \{u\}$.

**walk:** Sequence of vertices, with each vertex being adjacent to the vertices immediately preceding and succeeding it in the sequence.

# Problem Definitions

## $(d, l)$-CSP

Given a set of variables $V$, a domain $D$ of cardinality $d$, and a set of $l$-constraints $C$, that is relations between the values of the variables specifying the allowed combinations of values for a subset of size at most $l$ of variables, determine if there exists a satisfying assignment for the variables.

## $k$-COLORING

Given a graph $G$, determine if there is a coloring of $G$ with at most $k$ colors.

## $k$-SAT

Given a boolean formula in conjunctive normal form where each clause has at most $k$ literals, determine if there is an assignment of its variables such that the formula evaluates to true.

## #SAT

Given a boolean formula, determine the number of different assignments of its variables such that the formula evaluates to true.

## BANDWIDTH

Given a graph $G = (V, E)$ on $n$ vertices, find a linear arrangement $L : V \rightarrow \{1, \dots, n\}$ of its vertices such that the maximum stretch $\max_{\{u,v\} \in E}\{|L(u) - L(v)|\}$ of the edges is minimized.

## BINARY KNAPSACK

Given a knapsack of capacity $c > 0$ and $n$ items where each item $u$ has a value $v(u) > 0$ and a weight $w(u) > 0$, find a subset of items $S$ that fit into the knapsack, $\sum_{u \in S} w(u) \leq c$, and the total value, $\sum_{u \in S} v(u)$, is maximized.

## CSP

Given a set of variables $V$, a domain $D$, and a set of constraints $C$, that is relations between the values of the variables specifying the allowed combinations of values for a subset of variables, determine if there exists a satisfying assignment for the variables.

CHROMATIC NUMBER
Given a graph $G$, find the minimum number of colors needed for a coloring of $G$.

EXACT HITTING SET
Given a universe $\mathcal{U}$ of elements and a collection $\mathcal{S}$ of subsets of $\mathcal{U}$, determine if there exists a subset of elements in $\mathcal{U}$ such that each set of $\mathcal{S}$ contains exactly one of these elements.

FEEDBACK VERTEX SET
Given a graph $G$, find a feedback vertex set of $G$ of minimum cardinality.

GRAPH HOMOMORPHISM
Given two graphs $G$ and $H$ on at most $n$ vertices each, determine if there exists a mapping $\varphi : V(G) \rightarrow V(H)$ such that for every $x, y \in V(G)$, $uv \in E(G)$ implies $\varphi(u)\varphi(v) \in E(H)$.

HAMILTONIAN CYCLE
Given a graph $G$, determine if $G$ has a cycle that visits each vertex exactly once and returns to the starting vertex.

HAMILTONIAN PATH
Given a graph $G$, determine if $G$ has a path that visits each vertex exactly once.

MAX $k$-SAT
Given a boolean formula in conjunctive normal form where each clause has at most $k$ literals, find an assignment of its variables satisfying a maximum number of clauses.

MAX 2-CSP
Given a $(2, 2)$-CSP instance, find an assignment of the variables satisfying a maximum number of constraints.

MAX CUT
Given a graph $G = (V, E)$, find a subset of vertices $A \subseteq V$ with a maximum number of edges with one endpoint in $A$ and the other in $V \setminus A$.

MAXIMUM INDEPENDENT SET
Given a graph $G$, find an independent set of $G$ of maximum cardinality.

MINIMUM $d$-HITTING SET
Given a universe $\mathcal{U}$ of elements and a collection $\mathcal{S}$ of subsets of size at most $d$ of $\mathcal{U}$, find a minimum number of elements in $\mathcal{U}$ such that each set of $\mathcal{S}$ contains at least one of these elements.

MINIMUM DOMINATING SET
Given a graph $G$, find a dominating set of $G$ of minimum cardinality.

MINIMUM HITTING SET
Given a universe $\mathcal{U}$ of elements and a collection $\mathcal{S}$ of subsets of $\mathcal{U}$, find a minimum number of elements in $\mathcal{U}$ such that each set of $\mathcal{S}$ contains at least one of these elements.

## Minimum Independent Dominating Set
Given a graph $G$, find a set of vertices that is an independent set and a dominating set of $G$ of minimum cardinality.

## Minimum Set Cover
Given a universe $\mathcal{U}$ and a collection $\mathcal{S}$ of subsets of $\mathcal{U}$, find a minimum number of subsets in $\mathcal{S}$ such that their union is equal to $\mathcal{U}$.

## Minimum Vertex Cover
Given a graph $G$, find an vertex cover of $G$ of minimum cardinality.

## Quadratic Assignment
Given a set $P$ of $n$ facilities, a set $L$ of $n$ locations, a weight function $w : P \times P \to \mathbb{R}$, and a distance function $d : L \times L \to \mathbb{R}$, find an assignment $f : P \to L$ such that the cost $\sum_{a,b \in P} w(a,b) \cdot d(f(a), f(b))$ is minimized.

## SAT
Given a boolean formula, determine if there is an assignment of its variables such that the formula evaluates to true.

## Subgraph Isomorphism
Given two graphs $G_1$ and $G_2$ where $n$ is the number of vertices of $G_2$, determine whether $G_1$ is isomorphic to a subgraph of $G_2$.

## Subset Sum
Given a set of integers $S$, determine if there is a non–empty subset of $S$ that sums up to zero.

## Traveling Salesman
Given a set $\{1, \ldots, n\}$ of $n$ cities and the distance $d(i, j)$ between every two cities $i$ and $j$, find a tour visiting all cities with minimum total distance. A *tour* is a permutation of the cities starting and ending in city 1.

## Treewidth
Given a graph, determine its treewidth. The notion of treewidth is defined in Chapter 6.

# Bibliography

[AAC+04]    Gabriela Alexe, Sorin Alexe, Yves Crama, Stephan Foldes, Peter L. Hammer, and Bruno Simeone, *Consensus algorithms for the generation of all maximal bicliques*, Discrete Applied Mathematics **145** (2004), 11–21.

[ABF+02]    Jochen Alber, Hans L. Bodlaender, Henning Fernau, Ton Kloks, and Rolf Niedermeier, *Fixed parameter algorithms for dominating set and related problems on planar graphs*, Algorithmica **33** (2002), no. 4, 461–493.

[Abr96]    Jean-Raymond Abrial, *The B-book: Assigning programs to meanings*, Cambridge University Press, 1996.

[ACG+99]    Giorgio Ausiello, Pierluigi Crescenzi, Giorgio Gambosi, Viggo Kann, Alberto Marchetti-Spaccamela, and Marco Protasi, *Complexity and approximation : combinatorial optimization problems and their approximability properties*, Springer, 1999.

[AJ03]    Ola Angelsmark and Peter Jonsson, *Improved algorithms for counting solutions in constraint satisfaction problems*, Proceedings of the 9th International Conference on Principles and Practice of Constraint Programming (CP 2003), Lecture Notes in Computer Science, vol. 2833, Springer, Berlin, 2003, pp. 81–95.

[Alb02]    Jochen Alber, *Exact algorithms for NP-hard problems on networks: Design, analysis, and implementation*, Ph.D. thesis, Universität Tübingen, Germany, 2002.

[Ang05]    Ola Angelsmark, *Constructing algorithms for constraint satisfaction and related problems: Methods and applications*, Ph.D. thesis, Linköping University, Sweden, 2005.

[AT06]    Ola Angelsmark and Johan Thapper, *Partitioning based algorithms for some colouring problems*, Recent Advances in Constraints, Revised Selected and Invited Papers of the 10th Joint ERCIM/CoLogNET International Workshop on Constraint Solving and Constraint Logic Programming (CSCLP 2005), Lecture Notes in Computer Science, vol. 3978, Springer, Berlin, 2006, pp. 44–58.

[AVAD76]  Georgi Adel'son-Vel'skii, Vladimir Arlazarov, and Mikhail Donskoi, *Programirovanie igr (programming of games)*, Nauka, Moscow, 1976, in Russian.

[AVJ98]  Jérôme Amilhastre, Marie-Catherine Vilarem, and Philippe Janssen, *Complexity of minimum biclique cover and minimum biclique decomposition for bipartite domino-free graphs*, Discrete Applied Mathematics **86** (1998), 125–144.

[BBF99]  Vineet Bafna, Piotr Berman, and Toshihiro Fujito, *A 2-approximation algorithm for the undirected feedback vertex set problem*, SIAM Journal on Discrete Mathematics **12** (1999), no. 3, 289–297.

[BE95]  Richard Beigel and David Eppstein, *3-coloring in time $O(1.3446^n)$: A no-MIS algorithm*, Proceedings of the 36th Symposium on Foundations of Computer Science (FOCS 1995), IEEE, 1995, pp. 444–452.

[BE05]  ———, *3-coloring in time $O(1.3289^n)$*, Journal of Algorithms **54** (2005), no. 2, 168–204.

[BECF$^+$06]  Kevin Burrage, Vladimir Estivill-Castro, Michael R. Fellows, Michael A. Langston, Shev Mac, and Frances A. Rosamond, *The undirected feedback vertex set problem has a poly(k) kernel*, Proceedings of the 2nd International Workshop on Parameterized and Exact Computation (IWPEC 2006), Lecture Notes in Computer Science, vol. 4169, Springer, Berlin, 2006, pp. 192–202.

[Bei70]  Lowell W. Beineke, *Characterizations of derived graphs*, Journal of Combinatorial Theory, Series B **9** (1970), 129–135.

[Bei99]  Richard Beigel, *Finding maximum independent sets in sparse and general graphs*, Proceedings of the 10th ACM-SIAM Symposium on Discrete Algorithms (SODA 1999), ACM and SIAM, 1999, pp. 856–857.

[Ber84]  Alan A. Bertossi, *Dominating sets for split and bipartite graphs*, Information Processing Letters **19** (1984), 37–40.

[BFH94]  Hans L. Bodlaender, Michael R. Fellows, and Michael T. Hallett, *Beyond NP-completeness for problems of bounded width: Hardness for the W hierarchy*, Proceedings of the 26th Annual ACM Symposium on the Theory of Computing (STOC 1994), ACM, 1994, pp. 449–458.

[BH08]  Andreas Björklund and Thore Husfeldt, *Exact algorithms for exact satisfiability and number of perfect matchings*, Algorithmica **52** (2008), no. 2, 226–249.

[BHK09]  Andreas Björklund, Thore Husfeldt, and Mikko Koivisto, *Set partitioning via inclusion-exclusion*, SIAM Journal on Computing **39** (2009), no. 2, 546–563.

[BHKK07]  Andreas Björklund, Thore Husfeldt, Petteri Kaski, and Mikko Koivisto, *Fourier meets Möbius: fast subset convolution*, Proceedings of the 39th Annual ACM Symposium on Theory of Computing (STOC 2007), ACM, 2007, pp. 67–74.

[BHKK08a] _____, *Computing the tutte polynomial in vertex-exponential time*, Proceedings of the 49th Annual IEEE Symposium on Foundations of Computer Science (FOCS 2008), IEEE, 2008, pp. 677–686.

[BHKK08b] _____, *The travelling salesman problem in bounded degree graphs*, Proceedings of the 35th International Colloquium on Automata, Languages and Programming (ICALP 2008), Lecture Notes in Computer Science, vol. 5125, Springer, Berlin, 2008, pp. 198–209.

[BHKK08c] _____, *Trimmed moebius inversion and graphs of bounded degree*, Proceedings of the 25th Annual Symposium on Theoretical Aspects of Computer Science (STACS 2008), Dagstuhl Seminar Proceedings, vol. 08001, Internationales Begegnungs- und Forschungszentrum fuer Informatik (IBFI), Schloss Dagstuhl, Germany, 2008, pp. 85–96.

[BJ82] Kellogg S. Booth and J. Howard Johnson, *Dominating sets in chordal graphs*, SIAM Journal of Computing **11** (1982), no. 1, 191–199.

[Bjö07] Andreas Björklund, *Algorithmic bounds for presumably hard combinatorial problems*, Ph.D. thesis, Lund University, Sweden, 2007.

[BK04] Tobias Brueggemann and Walter Kern, *An improved deterministic local search algorithm for 3-SAT*, Theoretical Computer Science **329** (2004), no. 1-3, 303–313.

[BMS05] Jesper M. Byskov, Bolette A. Madsen, and Bjarke Skjernaa, *On the number of maximal bipartite subgraphs of a graph*, Journal of Graph Theory **48** (2005), no. 2, 127–132.

[BMT00] Lorenzo Brunetta, Francesco Maffioli, and Marco Trubian, *Solving the feedback vertex set problem on undirected graphs*, Discrete Applied Mathematics **101** (2000), no. 1-3, 37–51.

[Bod98] Hans L. Bodlaender, *A partial k-arboretum of graphs with bounded treewidth*, Theoretical Computer Science **209** (1998), no. 1-2, 1–45.

[Bod07] Hans L. Bodlaender, *A cubic kernel for feedback vertex set*, Proceedings of the 24th Annual Symposium on Theoretical Aspects of Computer Science (STACS 2007), Lecture Notes in Computer Science, vol. 4393, Springer, Berlin, 2007, pp. 320–331.

[BR99] Nikhil Bansal and Venkatesh Raman, *Upper bounds for MaxSat: Further improved*, Proceedings of the 10th International Symposium on Algorithms and Computation (ISAAC 1999), Lecture Notes in Computer Science, vol. 1741, Springer, Berlin, 1999, pp. 247–258.

[BT97] Hans L. Bodlaender and Dimitrios M. Thilikos, *Treewidth for graphs with small chordality*, Discrete Applied Mathematics **79** (1997), 45–61.

[BT01]    Vincent Bouchitté and Ioan Todinca, *Treewidth and minimum fill-in: grouping the minimal separators*, SIAM Journal on Computing **31** (2001), 212–232.

[BYGNR98] Reuven Bar-Yehuda, Dan Geiger, Joseph Naor, and Ron M. Roth, *Approximation algorithms for the feedback vertex set problem with applications to constraint satisfaction and Bayesian inference*, SIAM Journal on Computing **27** (1998), no. 4, 942–959.

[Bys04a]  Jesper M. Byskov, *Enumerating maximal independent sets with applications to graph colouring*, Operations Research Letters **32** (2004), no. 6, 547–556.

[Bys04b]  _____ , *Exact algorithms for graph colouring and exact satisfiability*, Ph.D. thesis, Aarhus University, Denmark, 2004.

[CB94]    Ramon Carbó and Emili Besalú, *Definition, mathematical examples and quantum chemical applications of nested summation symbols and logical kronecker deltas*, Computers & Chemistry **18** (1994), no. 2, 117–126.

[CDZ02]   Mao-Cheng Cai, Xiaotie Deng, and Wenan Zang, *A min-max theorem on feedback vertex sets*, Mathematics of Operations Research **27** (2002), no. 2, 361–371.

[CFL+08]  Jianer Chen, Fedor V. Fomin, Yang Liu, Songjian Lu, and Yngve Villanger, *Improved algorithms for feedback vertex set problems*, Journal of Computer and System Sciences **74** (2008), no. 7, 1188–1198.

[CGHW98]  Fabián A. Chudak, Michel X. Goemans, Dorit S. Hochbaum, and David P. Williamson, *A primal-dual interpretation of two 2-approximation algorithms for the feedback vertex set problem in undirected graphs*, Operations Research Letters **22** (1998), no. 4-5, 111–118.

[Chr71]   Nicos Christofides, *An algorithm for the chromatic number of a graph*, The Computer Journal **14** (1971), 38–39.

[CKJ01]   Jianer Chen, Iyad A. Kanj, and Weijia Jia, *Vertex cover: further observations and further improvements*, Journal of Algorithms **41** (2001), no. 2, 280–301.

[CKX05]   Jianer Chen, Iyad A. Kanj, and Ge Xia, *Labeled search trees and amortized analysis: improved upper bounds for NP-hard problems*, Algorithmica **43** (2005), no. 4, 245–273.

[CKX06]   _____ , *Improved parameterized upper bounds for vertex cover*, Proceedings of the 31st International Symposium on Mathematical Foundations of Computer Science (MFCS 2006), Lecture Notes in Computer Science, vol. 4162, Springer, Berlin, 2006, pp. 238–249.

[CLL+08]  Jianer Chen, Yang Liu, Songjian Lu, Barry O'Sullivan, and Igor Razgon, *A fixed-parameter algorithm for the directed feedback vertex set problem*, Proceedings of the 40th Annual ACM Symposium on Theory of Computing (STOC 2008), ACM, 2008, pp. 177–186.

[Coo71] Stephen A. Cook, *The complexity of theorem proving procedures*, Proceedings of the 3rd Annual ACM Symposium on Thoery of Computing (STOC 1971), ACM, 1971, pp. 151–158.

[CP08] Marek Cygan and Marcin Pilipczuk, *Faster exact bandwidth*, Proceedings of the 34th International Workshop on Graph-Theoretic Concepts in Computer Science (WG 2008), LNCS, vol. 5344, Springer, 2008, pp. 101–109.

[CP09a] _____, *Even faster exact bandwidth*, Tech. Report 0902.1661v1 [cs.CC], arXiv, 2009.

[CP09b] _____, *Exact and approximate bandwidth*, Proceedings of the 36th International Colloquium on Automata, Languages and Programming (ICALP 2009), LNCS, vol. 5555, Springer, 2009, pp. 304–315.

[DdFS05] Vânia M. F. Dias, Celina M. H. de Figueiredo, and Jayme L. Szwarcfiter, *Generating bicliques of a graph in lexicographic order*, Theoretical Computer Science **337** (2005), 240–248.

[DdFS07] _____, *On the generation of bicliques of a graph*, Discrete Applied Mathematics **155** (2007), 1826–1832.

[DF92] Rodney G. Downey and Michael R. Fellows, *Fixed-parameter intractability*, Proceedings of the Seventh Annual IEEE Structure in Complexity Theory Conference (SCT 1992), IEEE, 1992, pp. 36–49.

[DF99] _____, *Parameterized complexity*, Springer-Verlag, New York, 1999.

[DFL+05] Frank K. H. A. Dehne, Michael R. Fellows, Michael A. Langston, Frances A. Rosamond, and Kim Stevens, *An $O(2^{O(k)}n^3)$ FPT algorithm for the undirected feedback vertex set problem*, Proceedings of the 11th Annual International Conference on Computing and Combinatorics (COCOON 2005), Lecture Notes in Computer Science, vol. 3595, Springer, Berlin, 2005, pp. 859–869.

[DFL+07] _____, *An $O(2^{O(k)}n^3)$ FPT algorithm for the undirected feedback vertex set problem*, Theory of Computing Systems **41** (2007), no. 3, 479–492.

[DGH+02] Evgeny Dantsin, Andreas Goerdt, Edward A. Hirsch, Ravi Kannan, Jon Kleinberg, Christos Papadimitriou, Prabhakar Raghavan, and Uwe Schöning, *A deterministic $(2 - 2/(k + 1))^n$ algorithm for k-SAT based on local search*, Theoretical Computer Science **289** (2002), no. 1, 69–83.

[DGMS07] Erik D. Demaine, Gregory Gutin, Dániel Marx, and Ulrike Stege, *Open problems*, Structure Theory and FPT Algorithmes for Graphs, Digraphs and Hypergraphs, Dagstuhl Seminar Proceedings, vol. 07281, Internationales Begegnungs- und Forschungszentrum fuer Informatik (IBFI), Schloss Dagstuhl, Germany, 2007.

[DHIV01] Evgeny Dantsin, Edward A. Hirsch, Sergei Ivanov, and Maxim Vsemirnov, *Algorithms for SAT and upper bounds on their complexity*, Electronic Colloquium on Computational Complexity (ECCC) **8** (2001), no. 12.

[DHW06] Evgeny Dantsin, Edward A. Hirsch, and Alexander Wolpert, *Clause shortening combined with pruning yields a new upper bound for deterministic SAT algorithms*, Proceedings of the 6th Italian Conference on Algorithms and Complexity (CIAC 2006), Lecture Notes in Computer Science, vol. 3998, Springer, Berlin, 2006, pp. 60–68.

[DJ02] Vilhelm Dahllöf and Peter Jonsson, *An algorithm for counting maximum weighted independent sets and its applications*, Proceedings of the 13th Annual ACM-SIAM Symposium on Discrete Algorithms (SODA 2002), ACM and SIAM, 2002, pp. 292–298.

[DJW05] Vilhelm Dahllöf, Peter Jonsson, and Magnus Wahlström, *Counting models for 2SAT and 3SAT formulae*, Theoretical Computer Science **332** (2005), no. 1-3, 265–291.

[DKST01] Milind Dawande, Pinar Keskinocak, Jayashankar M. Swaminathand, and Sridhar Tayur, *On bipartite and multipartite clique problems*, Journal of Algorithms **41** (2001), 388–403.

[DLL+06] Frank K. H. A. Dehne, Michael A. Langston, Xuemei Luo, Sylvain Pitre, Peter Shaw, and Yun Zhang, *The cluster editing problem: Implementations and experiments*, Proceedings of the Second International Workshop on Parameterized and Exact Computation (IWPEC 2006), Lecture Notes in Computer Science, vol. 4169, Springer, Berlin, 2006, pp. 13–24.

[Dor06] Frederic Dorn, *Dynamic programming and fast matrix multiplication*, Proceedings of the 14th Annual European Symposium on Algorithms (ESA 2006), Lecture Notes in Computer Science, vol. 4168, Springer, Berlin, 2006, pp. 280–291.

[Dor07] ———, *Designing subexponential algorithms: Problems, techniques & structures*, Ph.D. thesis, University of Bergen, Norway, 2007.

[DP02] Limor Drori and David Peleg, *Faster exact solutions for some NP-hard problems*, Theoretical Computer Science **287** (2002), no. 2, 473–499.

[Edm65] Jack Edmonds, *Paths, trees, and flowers*, Canadian Journal of Mathematics **17** (1965), 449–467.

[EK72] Jack Edmonds and Richard M. Karp, *Theoretical improvements in algorithmic efficiency for network flow problems*, Journal of the ACM **19** (1972), no. 2, 248–264.

[ENSZ00] Guy Even, Joseph Naor, Baruch Schieber, and Leonid Zosin, *Approximating minimum subset feedback sets in undirected graphs with applications*, SIAM Journal on Discrete Mathematics **13** (2000), no. 2, 255–267.

[Epp03] David Eppstein, *Small maximal independent sets and faster exact graph coloring*, Journal of Graph Algorithms and Applications **7** (2003), no. 2, 131–140.

[Epp06] _____, *Quasiconvex analysis of multivariate recurrence equations for backtracking algorithms*, ACM Transactions on Algorithms **2** (2006), no. 4, 492–509.

[Fei00] Uriel Feige, *Coping with the NP-hardness of the graph bandwidth problem*, Proceedings of the 7th Scandinavian Workshop on Algorithm Theory (SWAT 2000), Lecture Notes in Computer Science, vol. 1851, Springer, Berlin, 2000, pp. 10–19.

[Fer06] Henning Fernau, *Parameterized algorithms for hitting set: The weighted case*, Proceedings of the 6th Italian Conference on Algorithms and Complexity (CIAC 2006), Lecture Notes in Computer Science, vol. 3998, Springer, Berlin, 2006, pp. 332–343.

[FG06] Jörg Flum and Martin Grohe, *Parameterized complexity theory*, Texts in Theoretical Computer Science. An EATCS Series, Springer, Berlin, 2006.

[FGK03] Robert Fourer, David M. Gay, and Brian W. Kernighan, *AMPL: A modeling language for mathematical programming*, second ed., Duxbury Press/Brooks/Cole Publishing Co., 2003.

[FGK05] Fedor V. Fomin, Fabrizio Grandoni, and Dieter Kratsch, *Some new techniques in design and analysis of exact (exponential) algorithms*, Bulletin of the EATCS **87** (2005), 47–77.

[FGK+07] Fedor V. Fomin, Petr A. Golovach, Jan Kratochvíl, Dieter Kratsch, and Mathieu Liedloff, *Branch and recharge: Exact algorithms for generalized domination*, Proceedings of the 10th International Workshop on Algorithms and Data Structures (WADS 2007), Lecture Notes in Computer Science, vol. 4619, Springer, Berlin, 2007, pp. 507–518.

[FGK+08] Fedor V. Fomin, Serge Gaspers, Dieter Kratsch, Mathieu Liedloff, and Saket Saurabh, *Iterative compression and exact algorithms*, Proceedings of the 33rd International Symposium on Mathematical Foundations of Computer Science (MFCS 2008), Lecture Notes in Computer Science, vol. 5162, Springer, Berlin, 2008, pp. 335–346.

[FGK+09a]   Henning Fernau, Serge Gaspers, Dieter Kratsch, Mathieu Liedloff, and Daniel Raible, *Exact exponential-time algorithms for finding bicliques in a graph*, Proceedings of the 8th Cologne-Twente Workshop on Graphs and Combinatorial Optimization (CTW 2009), Ecole Polytechnique and CNAM, 2009, pp. 205–209.

[FGK09b]   Fedor V. Fomin, Fabrizio Grandoni, and Dieter Kratsch, *A measure & conquer approach for the analysis of exact algorithms*, Journal of the ACM **56** (2009), no. 5, 1–32.

[FGK09c]   Martin Fürer, Serge Gaspers, and Shiva P. Kasiviswanathan, *An exponential time 2-approximation algorithm for bandwidth*, Proceedings of the 4th International Workshop on Parameterized and Exact Computation (IWPEC 2009), Lecture Notes in Computer Science, vol. 5917, Springer, 2009, to appear.

[FGP06]   Fedor V. Fomin, Serge Gaspers, and Artem V. Pyatkin, *Finding a minimum feedback vertex set in time $O(1.7548^n)$*, Proceedings of the 2nd International Workshop on Parameterized and Exact Computation (IWPEC 2006), Lecture Notes in Computer Science, vol. 4169, Springer, Berlin, 2006, pp. 184–191.

[FGPR08]   Fedor V. Fomin, Serge Gaspers, Artem V. Pyatkin, and Igor Razgon, *On the minimum feedback vertex set problem: Exact and enumeration algorithms*, Algorithmica **52** (2008), no. 2, 293–307.

[FGPS08]   Fedor V. Fomin, Fabrizio Grandoni, Artem V. Pyatkin, and Alexey A. Stepanov, *Combinatorial bounds via measure and conquer: Bounding minimal dominating sets and applications*, ACM Transactions on Algorithms **5** (2008), no. 1, 1–17.

[FGR09]   Henning Fernau, Serge Gaspers, and Daniel Raible, *Exact and parameterized algorithms for max internal spanning tree*, Proceedings of the 35th International Workshop on Graph-Theoretic Concepts in Computer Science (WG 2009), Lecture Notes in Computer Science, vol. 5911, Springer, Berlin, 2009, pp. 100–111.

[FGS06]   Fedor V. Fomin, Serge Gaspers, and Saket Saurabh, *Branching and treewidth based exact algorithms*, Proceedings of the 17th Annual International Symposium on Algorithms and Computation (ISAAC 2006), Lecture Notes in Computer Science, vol. 4288, Springer, Berlin, 2006, pp. 16–25.

[FGS07]   _____ , *Improved exact algorithms for counting 3- and 4-colorings*, Proceedings of the 13th Annual International Computing and Combinatorics Conference (COCOON 2007), Lecture Notes in Computer Science, vol. 4598, Springer, Berlin, 2007, pp. 65–74.

[FGSS09]   Fedor V. Fomin, Serge Gaspers, Saket Saurabh, and Alexey A. Stepanov, *On two techniques of combining branching and treewidth*, Algorithmica **54** (2009), no. 2, 181–207.

[FH77]     Stephan Földes and Peter L. Hammer, *Split graphs*, Proceedings of the 8th Southeastern Conference on Combinatorics, Graph Theory and Computing, 1977, pp. 311–315.

[FH06]     Fedor V. Fomin and Kjartan Høie, *Pathwidth of cubic graphs and exact algorithms*, Information Processing Letters **97** (2006), no. 5, 191–196.

[FK98]     Uriel Feige and Joe Kilian, *Zero knowledge and the chromatic number*, Journal of Computer and System Sciences **57** (1998), no. 2, 187–199.

[FK04]     Sergey S. Fedin and Alexander S. Kulikov, *Automated proofs of upper bounds on the running time of splitting algorithms*, Proceedings of the 1st International Workshop on Parameterized and Exact Computation (IWPEC 2004), Lecture Notes in Computer Science, vol. 3162, Springer, Belin, 2004, pp. 248–259.

[FK05]     Martin Fürer and Shiva P. Kasiviswanathan, *Algorithms for counting 2-SAT solutions and colorings with applications*, Tech. report, Electronic Colloquium on Computational Complexity (ECCC), 2005.

[FKW04]    Fedor V. Fomin, Dieter Kratsch, and Gerhard J. Woeginger, *Exact (exponential) algorithms for the dominating set problem*, Proceedings of the 30th Workshop on Graph Theoretic Concepts in Computer Science (WG 2004), Lecture Notes in Computer Science, vol. 3353, Springer, Berlin, 2004, pp. 245–256.

[FLRS07]   Michael R. Fellows, Michael A. Langston, Frances A. Rosamond, and Peter Shaw, *Efficient parameterized preprocessing for cluster editing*, Proceedings of the 16th International Symposium on Fundamentals of Computation Theory (FCT 2007), Lecture Notes in Computer Science, vol. 4639, Springer, Berlin, 2007, pp. 312–321.

[FP05]     Fedor V. Fomin and Artem V. Pyatkin, *Finding minimum feedback vertex set in bipartite graph*, Reports in Informatics 291, University of Bergen, 2005.

[FPR99]    Paola Festa, Panos M. Pardalos, and Mauricio G. C. Resende, *Feedback set problems*, Handbook of combinatorial optimization, Supplement Vol. A, Kluwer Academic Publishers, Dordrecht, 1999, pp. 209–258.

[FR96]     Meinrad Funke and Gerhard Reinelt, *A polyhedral approach to the feedback vertex set problem*, Integer programming and combinatorial optimization, Lecture Notes in Computer Science, vol. 1084, Springer, Berlin, 1996, pp. 445–459.

[Fuk96]    Komei Fukuda, *Complexity of enumeration - evaluating the hardness of listing objects*, presented at ETH Zurich, May 1996, also at International Symposium on Mathematical Programming 1997, 1996, http://www.ifor.math.ethz.ch/ ~fukuda/old/ENP_home/ENP_note.html.

[FV08]     Fedor V. Fomin and Yngve Villanger, *Treewidth computation and extremal combinatorics*, Proceedings of the 35th International Colloquium on Automata, Languages and Programming (ICALP 2008), Lecture Notes in Computer Science, vol. 5125, Springer, Berlin, 2008, pp. 210–221.

[FV10]     _____, *Finding induced subgraphs via minimal triangulations*, Proceedings of the 27th International Symposium on Theoretical Aspects of Computer Science (STACS 2010), Leibniz International Proceedings in Informatics, Schloss Dagstuhl - Leibniz Center of Informatics, 2010, to appear.

[Gas02]    William I. Gasarch, *Guest column: The P=?NP poll*, SIGACT News **33** (2002), no. 2, 34–47.

[Gas05]    Serge Gaspers, *Algorithmes exponentiels*, Master's thesis, University of Metz, France, 2005, in French.

[Gas08]    Serge Gaspers, *Exponential time algorithms: Structures, measures, and bounds*, Ph.D. thesis, University of Bergen, Norway, 2008.

[Gas09]    Serge Gaspers, *Measure & conquer for parameterized branching algorithms*, Parameterized Complexity News: Newsletter of the Parameterized Complexity Community, September 2009.

[GGH+06]   Jiong Guo, Jens Gramm, Falk Hüffner, Rolf Niedermeier, and Sebastian Wernicke, *Compression-based fixed-parameter algorithms for feedback vertex set and edge bipartization*, Journal of Computer and System Sciences **72** (2006), no. 8, 1386–1396.

[GHNR03]   Jens Gramm, Edward A. Hirsch, Rolf Niedermeier, and Peter Rossmanith, *Worst-case upper bounds for MAX-2-SAT with an application to MAX-CUT*, Discrete Applied Mathematics **130** (2003), no. 2, 139–155.

[GJ79]     Michael R. Garey and David S. Johnson, *Computers and intractability, a guide to the theory of NP-completeness*, W.H. Freeman and Company, New York, 1979.

[GKL06]    Serge Gaspers, Dieter Kratsch, and Mathieu Liedloff, *Exponential time algorithms for the minimum dominating set problem on some graph classes*, Proceedings of the 10th Scandinavian Workshop on Algorithm Theory (SWAT 2006), Lecture Notes in Computer Science, vol. 4059, Springer, Berlin, 2006, pp. 148–159.

[GKL08]    _____, *On independent sets and bicliques in graphs*, Proceedings of the 34th International Workshop on Graph-Theoretic Concepts in Computer Science (WG 2008), Lecture Notes in Computer Science, vol. 5344, Springer, Berlin, 2008, pp. 171–182.

[GKLT09]  Serge Gaspers, Dieter Kratsch, Mathieu Liedloff, and Ioan Todinca, *Exponential time algorithms for the minimum dominating set problem on some graph classes*, ACM Transactions on Algorithms **6** (2009), no. 1:9, 1–21.

[GL06]    Serge Gaspers and Mathieu Liedloff, *A branch-and-reduce algorithm for finding a minimum independent dominating set in graphs*, Proceedings of the 32nd International Workshop on Graph-Theoretic Concepts in Computer Science (WG 2006), Lecture Notes in Computer Science, vol. 4271, Springer, Berlin, 2006, pp. 78–89.

[GM09]    Serge Gaspers and Matthias Mnich, *On feedback vertex sets in tournaments*, Tech. Report 0905.0567v1 [cs.DM], arXiv, 2009.

[Gol78]   Martin C. Golumbic, *Trivially perfect graphs*, Discrete Mathematics **24** (1978), 105–107.

[Gol80]   ———, *Algorithmic graph theory and perfect graphs*, Academic Press, New York, 1980.

[Gra04]   Fabrizio Grandoni, *Exact algorithms for hard graph problems*, Ph.D. thesis, Università di Roma "Tor Vergata", Roma, Italy, 2004.

[Gra06]   ———, *A note on the complexity of minimum dominating set*, Journal of Discrete Algorithms **4** (2006), no. 2, 209–214.

[GRS06]   Sushmita Gupta, Venkatesh Raman, and Saket Saurabh, *Fast exponential algorithms for maximum r-regular induced subgraph problems*, Proceedings of the 26th Conference on Foundations of Software Technology and Theoretical Computer Science (FSTTCS 2006), Lecture Notes in Computer Science, no. 4337, Springer, Berlin, 2006, pp. 139–151.

[GS87]    Yuri Gurevich and Saharon Shelah, *Expected computation time for Hamiltonian path problem*, SIAM Journal on Computing **16** (1987), no. 3, 486–502.

[GS09]    Serge Gaspers and Gregory Sorkin, *A universally fastest algorithm for Max 2-Sat, Max 2-CSP, and everything in between*, Proceedings of the 20th Annual ACM-SIAM Symposium on Discrete Algorithms (SODA 2009), ACM and SIAM, 2009, pp. 606–615.

[GSB95]   Mark K. Goldberg, Thomas H. Spencer, and David A. Berque, *A low-exponential algorithm for counting vertex covers*, Graph Theory, Combinatorics, Algorithms, and Applications, vol. 1, Wiley, New York, 1995, pp. 431–444.

[Guo09]   Jiong Guo, *A more effective linear kernelization for cluster editing*, Theoretical Computer Science **410** (2009), no. 8-10, 718–726.

[GW96]    Bernhard Ganter and Rudolf Wille, *Formal concept analysis, mathematical foundations*, Springer, Berlin, 1996.

[Hir00]   Edward A. Hirsch, *A new algorithm for MAX-2-SAT*, Proceedings of the 17th Annual Symposium on Theoretical Aspects of Computer Science (STACS 2000), Lecture Notes in Computer Science, vol. 1770, Springer, Berlin, 2000, pp. 65–73.

[HK62]    Michael Held and Richard M. Karp, *A dynamic programming approach to sequencing problems*, Journal of SIAM **10** (1962), 196–210.

[HK73]    John E. Hopcroft and Richard M. Karp, *An $n^{5/2}$ algorithm for maximum matching in bipartite graphs*, SIAM Journal on Computing **2** (1973), no. 4, 225–231.

[HKMN08]  Falk Hüffner, Christian Komusiewicz, Hannes Moser, and Rolf Niedermeier, *Fixed-parameter algorithms for cluster vertex deletion*, Proceedings of the 8th Latin American Theoretical Informatics Symposium (LATIN 2008), Lecture Notes in Computer Science, vol. 4957, Springer, Berlin, 2008, pp. 711–722.

[Hoc98]   Dorit S. Hochbaum, *Approximating clique and biclique problems*, Journal of Algorithms **29** (1998), 174–200.

[HS74]    Ellis Horowitz and Sartaj Sahni, *Computing partitions with applications to the knapsack problem*, Journal of the ACM **21** (1974), no. 2, 277–292.

[HT93]    Mihály Hujter and Zsolt Tuza, *The number of maximal independent sets in triangle-free graphs*, SIAM Journal on Discrete Mathematics **6** (1993), 284–288.

[IP99]    Russell Impagliazzo and Ramamohan Paturi, *Complexity of k-SAT*, Proceedings of the 14th Annual IEEE Conference on Computational Complexity, IEEE, 1999, pp. 237–240.

[IPZ01]   Russell Impagliazzo, Ramamohan Paturi, and Francis Zane, *Which problems have strongly exponential complexity*, Journal of Computer and System Sciences **63** (2001), no. 4, 512–530.

[Iwa04]   Kazuo Iwama, *Worst-case upper bounds for k-SAT*, Bulletin of the EATCS **82** (2004), 61–71.

[Jia86]   Tang Jian, *An $O(2^{0.304n})$ algorithm for solving maximum independent set problem*, IEEE Transactions on Computers **35** (1986), no. 9, 847–851.

[JRR03]   Michael Jünger, Gerhard Reinelt, and Giovanni Rinaldi, *Preface*, Combinatorial Optimization - Eureka, you shrink!, Lecture Notes in Computer Science, vol. 2570, Springer, Berlin, 2003.

[JS99]    David S. Johnson and Mario Szegedy, *What are the least tractable instances of max independent set?*, Proceedings of the 10th ACM-SIAM Symposium on Discrete Algorithms (SODA 1999), ACM and SIAM, 1999, pp. 927–928.

[JYP88]     David S. Johnson, Mihalis Yannakakis, and Christos H. Papadimitriou, *On generating all maximal independent sets*, Information Processing Letters **27** (1988), no. 3, 119–123.

[Kar72]     Richard M. Karp, *Reducibility among combinatorial problems*, Complexity of computer computations, Plenum Press, New York, 1972, pp. 85–103.

[Kar82]     ———, *Dynamic programming meets the principle of inclusion and exclusion*, Operations Research Letters **1** (1982), no. 2, 49–51.

[Kei93]     J. Mark Keil, *The complexity of domination problems in circle graphs*, Discrete Applied Mathematics **42** (1993), no. 1, 51–63.

[KK01]      Jon Kleinberg and Amit Kumar, *Wavelength conversion in optical networks*, Journal of Algorithms **38** (2001), no. 1, 25–50.

[KK06]      Arist Kojevnikov and Alexander S. Kulikov, *A new approach to proving upper bounds for MAX-2-SAT*, Proceedings of the 17th Annual ACM–SIAM Symposium on Discrete Algorithms (SODA 2006), ACM and SIAM, New York, 2006, pp. 11–17.

[KK07]      Alexander S. Kulikov and Konstantin Kutzkov, *New bounds for MAX-SAT by clause learning*, Proceedings of the 2nd International Symposium on Computer Science in Russia (CSR 2007), Lecture Notes in Computer Science, vol. 4649, Springer, Berlin, 2007, pp. 194–204.

[Klo94]     Ton Kloks, *Treewidth, computations and approximations*, Lecture Notes in Computer Science, vol. 842, Springer, 1994.

[Klo96]     ———, *Treewidth of circle graphs*, International Journal of Foundations of Computer Science **7** (1996), 111–120.

[KLR09]     Joachim Kneis, Alexander Langer, and Peter Rossmanith, *A fine-grained analysis of a simple independent set algorithm*, Proceedings of the 29th International Conference on Foundations of Software Technology and Theoretical Computer Science (FSTTCS 2009), Leibniz International Proceedings in Informatics, Schloss Dagstuhl - Leibniz Center of Informatics, 2009, to appear.

[KMRR09]    Joachim Kneis, Daniel Mölle, Stefan Richter, and Peter Rossmanith, *A bound on the pathwidth of sparse graphs with applications to exact algorithms*, SIAM Journal on Discrete Mathematics **23** (2009), no. 1, 407–427.

[Koi06]     Mikko Koivisto, *Optimal 2-constraint satisfaction via sum-product algorithms*, Information Processing Letters **98** (2006), no. 1, 24–28.

[KP06]      Subhash Khot and Ashok K. Ponnuswami, *Better inapproximability results for MaxClique, Chromatic Number and Min-3Lin-Deletion*, Proceedings of the 33rd

International Colloquium on Automata, Languages and Programming (ICALP 2006), Lecture Notes in Computer Science, vol. 4051, Springer, Berlin, 2006, pp. 226–237.

[KPS08]     Graham Kendall, Andrew Parkes, and Kristian Spoerer, *A survey of NP-complete puzzles*, International Computer Games Association Journal **31** (2008), no. 1, 13–34.

[KR05]      Joachim Kneis and Peter Rossmanith, *A new satisfiability algorithm with applications to Max-Cut*, Tech. Report AIB-2005-08, Department of Computer Science, RWTH Aachen, 2005.

[Kra07]     Dieter Kratsch, *In: Open problem session*, presented at Dagstuhl Seminar 07211 on Exact, Approximative, Robust and Certifying Algorithms on Particular Graph Classes, May 2007.

[KS93]      Ephraim Korach and Nir Solel, *Tree-width, path-width, and cutwidth*, Discrete Applied Mathematics **43** (1993), no. 1, 97–101.

[Kul99]     Oliver Kullmann, *New methods for 3-SAT decision and worst-case analysis*, Theoretical Computer Science **223** (1999), no. 1–2, 1–72.

[Kul05]     Alexander S. Kulikov, *Automated generation of simplification rules for SAT and MAXSAT*, Proceedings of the 8th International Conference on Theory and Applications of Satisfiability Testing (SAT 2005), Lecture Notes in Computer Science, vol. 3569, Springer, Berlin, 2005, pp. 430–436.

[KVZ01]     Andrew B. Kahng, Shailesh Vaya, and Alexander Zelikovsky, *New graph bipartizations for double-exposure, bright field alternating phase-shift mask layout*, Proceedings of the 6th Asia and South Pacific Design Automation Conference (ASP-DAC 2001), ACM, New York, 2001, pp. 133–138.

[KW05]      Bettina Klinz and Gerhard J. Woeginger, *Faster algorithms for computing power indices in weighted voting games*, Mathematical Social Sciences **49** (2005), no. 1, 111–116.

[Law76]     Eugene L. Lawler, *A note on the complexity of the chromatic number problem*, Information Processing Letters **5** (1976), no. 3, 66–67.

[Lev73]     Leonid A. Levin, *Universal search problems*, Problemy Peredaci Informacii **9** (1973), 265–266, In Russian; English translation in: Boris A. Trakhtenbrot, *A survey of russian approaches to perebor (brute-force search) algorithms*, Annals of the History of Computing **6** (1984), no. 4, 384–400.

[Lie07]     Mathieu Liedloff, *Algorithmes exacts et exponentiels pour les problèmes NP-difficiles : domination, variantes et généralisations*, Ph.D. thesis, University of Metz, France, 2007, in French.

[Lie08] ———, *Finding a dominating set on bipartite graphs*, Information Processing Letters **107** (2008), 154–157.

[LS09] Daniel Lokshtanov and Saket Saurabh, *Even faster algorithm for set splitting!*, Proceedings of the 4th International Workshop on Parameterized and Exact Computation (IWPEC 2009), Lecture Notes in Computer Science, vol. 5917, Springer, Berlin, 2009, pp. 288–299.

[MM65] John W. Moon and Leo Moser, *On cliques in graphs*, Israel Journal of Mathematics **3** (1965), 23–28.

[Moo71] John W. Moon, *On maximal transitive subtournaments*, Proceedings of the Edinburgh Mathematical Society **17** (1971), no. 4, 345–349.

[MP06] Burkhard Monien and Robert Preis, *Upper bounds on the bisection width of 3- and 4-regular graphs*, Journal of Discrete Algorithms **4** (2006), no. 3, 475–498.

[MS85] Burkhard Monien and Ewald Speckenmeyer, *Solving satisfiability in less than $2^n$ steps*, Discrete Applied Mathematics **10** (1985), no. 3, 287–295.

[MU04] Kazuhisa Makino and Takeaki Uno, *New algorithms for enumerating all maximal cliques*, Proceedings of the 9th Scandinavian Workshop on Algorithm Theory (SWAT 2004), Lecture Notes in Computer Science, vol. 3111, Springer, Berlin, 2004, pp. 260–272.

[Ned09] Jesper Nederlof, *Fast polynomial-space algorithms using möbius inversion: Improving on steiner tree and related problems*, Proceedings of the 36th International Colloquium on Automata, Languages and Programming (ICALP 2009), Lecture Notes in Computer Science, vol. 5555, Springer, Berlin, 2009, pp. 713–725.

[Nie06] Rolf Niedermeier, *Invitation to fixed-parameter algorithms*, Oxford Lecture Series in Mathematics and its Applications, vol. 31, Oxford University Press, Oxford, 2006.

[NR99] Lhouari Nourine and Olivier Raynaud, *A fast algorithm for building lattices*, Information Processing Letters **71** (1999), 199–204.

[NR00] Rolf Niedermeier and Peter Rossmanith, *New upper bounds for maximum satisfiability*, Journal of Algorithms **36** (2000), no. 1, 63–88.

[NR02] Lhouari Nourine and Olivier Raynaud, *A fast incremental algorithm for building lattices*, Journal of Experimental and Theoretical Artificial Intelligence **14** (2002), 217–227.

[OUU08] Yoshio Okamoto, Takeaki Uno, and Ryuhei Uehara, *Counting the number of independent sets in chordal graphs*, Journal of Discrete Algorithms **6** (2008), no. 2, 229–242.

[Pee03]    René Peeters, *The maximum edge biclique problem is NP-complete*, Discrete Applied Mathematics **131** (2003), 651–654.

[PI00]     Pavel Pudlák and Russel Impaglazzio, *A lower bound for DLL algorithms for k-SAT*, Proceedings of the 11th ACM-SIAM Symposium on Discrete Algorithms (SODA 2000), ACM and SIAM, 2000, pp. 128–136.

[PKS04]    Mihai Pop, Daniel S. Kosack, and Steven L. Salzberg, *Hierarchical scaffolding with bambus*, Genome Research **14** (2004), 149–159.

[PQR99]    Panos M. Pardalos, Tianbing Qian, and Mauricio G. C. Resende, *A greedy randomized adaptive search procedure for the feedback vertex set problem*, Journal of Combinatorial Optimization **2** (1999), no. 4, 399–412.

[Pri00]    Erich Prisner, *Bicliques in graphs I: Bounds on their number*, Combinatorica **20** (2000), 109–117.

[RF08]     Daniel Raible and Henning Fernau, *A new upper bound for Max-2-SAT: A graph-theoretic approach*, Proceedings of the 33rd International Symposium on Mathematical Foundations of Computer Science 2008 (MFCS 2008), Lecture Notes in Computer Science, vol. 5162, Springer, Berlin, 2008, pp. 551–562.

[Rie06]    Tobias Riege, *The domatic number problem: Boolean hierarchy completeness and exact exponential-time algorithms*, Ph.D. thesis, Heinrich-Heine University Düsseldorf, Germany, 2006.

[Rob86]    John M. Robson, *Algorithms for maximum independent sets*, Journal of Algorithms **7** (1986), no. 3, 425–440.

[Rob01]    _____, *Finding a maximum independent set in time $O(2^{n/4})$*, manuscript, http://www.labri.fr/perso/robson/mis/techrep.ps, 2001.

[RS83]     Neil Robertson and Paul D. Seymour, *Graph minors. I. Excluding a forest*, Journal of Combinatorial Theory, Series B **35** (1983), 39–61.

[RS86]     _____, *Graph minors. II. Algorithmic aspects of tree-width*, Journal of Algorithms **7** (1986), no. 3, 309–322.

[RS04]     Bert Randerath and Ingo Schiermeyer, *Exact algorithms for MINIMUM DOMINATING SET*, Technical Report zaik-469, Zentrum für Angewandte Informatik Köln, Germany, 2004.

[RSS05]    Venkatesh Raman, Saket Saurabh, and Somnath Sikdar, *Improved exact exponential algorithms for vertex bipartization and other problems*, Proceedings of the 9th Italian Conference on Theoretical Computer Science (ICTCS 2005), Lecture Notes in Computer Science, vol. 3701, Springer, Berlin, 2005, pp. 375–389.

[RSS06]   Venkatesh Raman, Saket Saurabh, and C. R. Subramanian, *Faster fixed param-eter tractable algorithms for finding feedback vertex sets*, ACM Transactions on Algorithms **2** (2006), no. 3, 403–415.

[RSS07]   Venkatesh Raman, Saket Saurabh, and Somnath Sikdar, *Efficient exact algorithms through enumerating maximal independent sets and other techniques*, Theory of Computing Systems **41** (2007), no. 3, 563–587.

[RSV04]   Bruce A. Reed, Kaleigh Smith, and Adrian Vetta, *Finding odd cycle transversals*, Operations Research Letters **32** (2004), no. 4, 299–301.

[RWB⁺07]  Sven Rahmann, Tobias Wittkop, Jan Baumbach, Marcel Martin, Anke Truss, and Sebastian Böcker, *Exact and heuristic algorithms for weighted cluster editing*, Proceedings of the 2005 IEEE Computational Systems Bioinformatics Conference **6** (2007), no. 1, 391–401.

[Sax80]   James B. Saxe, *Dynamic-programming algorithms for recognizing small-bandwidth graphs in polynomial time*, SIAM Journal on Algebraic and Discrete Methods **1** (1980), no. 4, 363–369.

[Sch95]   Ingo Schiermeyer, *Problems remaining NP-complete for sparse or dense graphs*, Discussiones Mathematicae. Graph Theory **15** (1995), 33–41.

[Sch01]   Uwe Schöning, *New algorithms for k-SAT based on the local search principle*, Pro-ceedings of the 26th International Symposium on Mathematical Foundations of Computer Science (MFCS 2001), Lecture Notes in Computer Science, vol. 2136, Springer, Berlin, 2001, pp. 87–95.

[Sch05]   ———, *Algorithmics in exponential time*, Proceedings of the 22nd International Symposium on Theoretical Aspects of Computer Science (STACS 2005), Lecture Notes in Computer Science, vol. 3404, Springer, Berlin, 2005, pp. 36–43.

[Sch08]   Dominik Scheder, *Guided search and a faster deterministic algorithm for 3-SAT*, Proceedings of the 8th Latin American Symposium on Theoretical Informatics (LATIN 2008), Lecture Notes in Computer Science, vol. 4957, Springer, Berlin, 2008, pp. 60–71.

[SS81]    Richard Schroeppel and Adi Shamir, *A $T = O(2^{n/2}), S = O(2^{n/4})$ algorithm for certain NP-complete problems*, SIAM Journal on Computing **10** (1981), no. 3, 456–464.

[SS02]    Benno Schwikowski and Ewald Speckenmeyer, *On enumerating all minimal so-lutions of feedback problems*, Discrete Applied Mathematics **117** (2002), no. 1-3, 253–265.

[SS03]     Alexander D. Scott and Gregory B. Sorkin, *Faster algorithms for MAX CUT and MAX CSP, with polynomial expected time for sparse instances*, Proceedings of the 7th International Workshop on Randomization and Approximation Techniques in Computer Science (RANDOM 2003), Lecture Notes in Computer Science, vol. 2764, Springer, Berlin, 2003, pp. 382–395.

[SS04]     ———, *A faster exponential-time algorithm for Max 2-Sat, Max Cut, and Max k-Cut*, Tech. Report RC23456 (W0412-001), IBM Research Report, December 2004.

[SS07]     ———, *Linear-programming design and analysis of fast algorithms for Max 2-CSP*, Discrete Optimization **4** (2007), no. 3–4, 260–287.

[SS09]     ———, *Polynomial constraint satisfaction problems, graph bisection, and the Ising partition function*, ACM Transactions on Algorithms **5** (2009), no. 4, 1–27.

[ST90]     Miklo Shindo and Etsuji Tomita, *A simple algorithm for finding a maximum clique and its worst-case time complexity*, Systems and Computers in Japan **21** (1990), no. 3, 1–13.

[Ste08]    Alexey A. Stepanov, *Exact algorithms for hard listing, counting and decision problems*, Ph.D. thesis, University of Bergen, Norway, 2008.

[Tho09]    Stéphan Thomassé, *A quadratic kernel for feedback vertex set*, Proceedings of the 20th Annual ACM-SIAM Symposium on Discrete Algorithms (SODA 2009), SIAM, 2009, pp. 115–119.

[Tra84]    Boris A. Trakhtenbrot, *A survey of russian approaches to perebor (brute-force search) algorithms*, Annals of the History of Computing, vol. 6, IEEE, 1984, pp. 384–400.

[Tra08]    Patrick Traxler, *The time complexity of constraint satisfaction*, Proceedings of the 3rd International Workshop on Parameterized and Exact Computation (IWPEC 2008), Lecture Notes in Computer Science, vol. 5018, Springer, Berlin, 2008, pp. 190–201.

[TT77]     Robert E. Tarjan and Anthony E. Trojanowski, *Finding a maximum independent set*, SIAM Journal on Computing **6** (1977), no. 3, 537–546.

[Ung98]    Walter Unger, *The complexity of the approximation of the bandwidth problem*, Proceedings of the 39th Annual Symposium on Foundations of Computer Science (FOCS 1998), IEEE, 1998, pp. 82–91.

[Val79]    Leslie G. Valiant, *The complexity of computing the permanent*, Theoretical Computer Science **8** (1979), no. 2, 189–201.

[vRB08a] Johan M. M. van Rooij and Hans L. Bodlaender, *Design by measure and conquer, a faster exact algorithm for dominating set*, Proceedings of the 25th Annual Symposium on Theoretical Aspects of Computer Science (STACS 2008), Dagstuhl Seminar Proceedings, vol. 08001, Internationales Begegnungs- und Forschungszentrum fuer Informatik (IBFI), 2008, pp. 657–668.

[vRB08b] ———, *Exact algorithms for edge domination*, Proceedings of the 3rd International Workshop on Parameterized and Exact Computation (IWPEC 2008), Lecture Notes in Computer Science, vol. 5018, Springer, Berlin, 2008, pp. 214–225.

[vRBR09] Johan M. M. van Rooij, Hans L. Bodlaender, and Peter Rossmanith, *Dynamic programming on tree decompositions using generalised fast subset convolution*, Proceedings of the 17th Annual European Symposium on Algorithms (ESA 2009), Lecture Notes in Computer Science, vol. 5757, Springer, Berlin, 2009, pp. 566–577.

[vRNvD09] Johan M. M. van Rooij, Jesper Nederlof, and Thomas C. van Dijk, *Inclusion/exclusion meets measure and conquer: Exact algorithms for counting dominating sets*, Proceedings of the 17th Annual European Symposium on Algorithms (ESA 2009), LNCS, vol. 5757, Springer, 2009, pp. 554–565.

[VWW06] Virginia Vassilevska, Ryan Williams, and Shan Leung Maverick Woo, *Confronting hardness using a hybrid approach*, Proceedings of the 17th Annual ACM-SIAM Symposium on Discrete Algorithms (SODA 2006), ACM, 2006, pp. 1–10.

[Wah04] Magnus Wahlström, *Exact algorithms for finding minimum transversals in rank-3 hypergraphs*, Journal of Algorithms **51** (2004), no. 2, 107–121.

[Wah07] ———, *Algorithms, measures and upper bounds for satisfiability and related problems*, Ph.D. thesis, Linköping University, Sweden, 2007.

[Wah08] ———, *A tighter bound for counting max-weight solutions to 2SAT instances*, Proceedings of the 3rd International Workshop on Parameterized and Exact Computation (IWPEC 2008), Lecture Notes in Computer Science, vol. 5018, Springer, Berlin, 2008, pp. 202–213.

[Wil05] Ryan Williams, *A new algorithm for optimal 2-constraint satisfaction and its implications*, Theoretical Computer Science **348** (2005), no. 2-3, 357–365.

[Wil07] ———, *Algorithms and resource requirements for fundamental problems*, Ph.D. thesis, Carnegie Mellon University, USA, 2007.

[Woe03] Gerhard J. Woeginger, *Exact algorithms for NP-hard problems: A survey*, Combinatorial Optimization - Eureka, you shrink!, Lecture Notes in Computer Science, vol. 2570, Springer, Berlin, 2003, pp. 185–207.

[Woe04]        _____, *Space and time complexity of exact algorithms: Some open problems*, Proceedings of the 1st International Workshop on Parameterized and Exact Computation (IWPEC 2004), Lecture Notes in Computer Science, vol. 3162, Springer, Berlin, 2004, pp. 281–290.

[Woe08]        _____, *Open problems around exact algorithms*, Discrete Applied Mathematics **156** (2008), 397–405.

[Yan78]        Mihalis Yannakakis, *Node and edge deletion NP-complete problems*, Proceedings of the 10th Annual ACM Symposium on Theory of Computing (STOC 1978), ACM, 1978, pp. 253–264.

[Zha96]        Wenhui Zhang, *Number of models and satisfiability of sets of clauses*, Theoretical Computer Science **155** (1996), no. 1, 277–288.

[Zuc07]        David Zuckerman, *Linear degree extractors and the inapproximability of max clique and chromatic number*, Theory of Computing **3** (2007), no. 1, 103–128.

# Index